THE HUMAN INJURY
A SPECIES IN CRISIS

by Anthony A. Wall, Jr.

DORRANCE
PUBLISHING CO
EST. 1920
PITTSBURGH, PENNSYLVANIA 15238

The contents of this work, including, but not limited to, the accuracy of events, people, and places depicted; opinions expressed; permission to use previously published materials included; and any advice given or actions advocated are solely the responsibility of the author, who assumes all liability for said work and indemnifies the publisher against any claims stemming from publication of the work.

Dorrance Publishing Co
585 Alpha Drive
Suite 103
Pittsburgh, PA 15238
Visit our website at www.dorrancebookstore.com

ISBN: 979-8-89649-179-8
eISBN: 979-8-89649-678-6

Book Dedication

To my beloved sons, Matthew, and John. I leave a template for Hope.

If I should leave while you're still here,
Know that I live on.
Vibrating to a different measure.
You will not see me,
So you must have faith.
Whenever you need me,
Just whisper my name in your heart;
I will be there.

—-*Ascension*, Colleen Hitchcock

—To my father, Anthony A. Wall, Sr. A misunderstood masterpiece.

—To Kathy McGee Burns and Mike Burns. Mom and Dad. Poppy and Grammy.

"The race is not given to the swift, but to those who endure." —Ecclesiastes 9:12

—To Laura DeVries. The Executive Producer. This is all your fault.

THE HUMAN INJURY
A SPECIES IN CRISIS

Foreword To The Human Injury

As Facebook readers often do, I read posts that are of interest because they take a novel look at rare topics of extreme interest. One of Tony Wall's posts popped up and I "liked" it, thus ensuring that more posts would arrive. Even though we were not yet friends, we were connected in other ways.

We both attended William Penn Charter School, Philadelphia. But Tony was one year behind me, and I didn't know him. In any event, I had not seen him or communicated with him for four decades. Tony's posts dealt with something that I had not read about before and had not even considered in detail—The Human Condition.

As I read Tony's posts, I became fascinated. He discussed what very few among us ever would—the totality of human degradation—but his observations never devolved into blame or judgment. His observations always come through the lens of empathy and compassion. He deftly discusses cosmological phenomena and other fascinating mathematical concepts. He also discussed why humanity is headed for extinction: not by its choices, but by its wounds. That's no small claim. He also spoke with equal passion about a responsible and scientific template for a sentient but troubled humanity to preserve itself. This was not sensationalism; this was science.

I sent Tony a friend request, which he graciously accepted. I began respectfully challenging some of his ideas and he always had a thoughtful, direct, and credible response to all my probing questions. I was uniquely impressed.

I was so fascinated, in fact, that I requested to personally meet with Tony. His ideas were so different, so I asked Tony if he would consider having coffee with me so that I might personally discuss his ideas with him and see if he could respond as confidently to my pointed questions in person. Tony agreed. I was due to be in the Philadelphia area on unrelated business, so I extended to him an invitation to meet. I was so glad that I did.

I found Tony to be as fascinating in person as in print. He's a man on a sincere mission to redirect an off-course humanity. Such a large undertaking seems incredible. Consider the missions of nonprofit organizations such as saving whales, saving national parks, preserving historical buildings, feeding the homeless, honoring veterans, etc. Those are eminently noble and important. But stopping the degradation of modern humanity? Now that is about the largest mission one could undertake, and yet here was Tony talking both urgently and cogently about just that.

Tony is 100% authentic and infectiously sincere. He's no mere dreamer. He's a do-er. I encouraged him to do more than just post about these ideas. No one was discussing anything close to the issues that concerned him. Tony had by then initiated the creation of an actionable non-profit. He followed through in full. It is called Noēsis. In fact, I became Tony's attorney to help him establish and register Noēsis and its trademark. I provided legal advice and counsel. I then watched Noēsis grow from an idea to a reality. It's grown now into an actionable vehicle of influence. I should add here that Noēsis existed well before the COVID-19 pandemic.

As I now view the growing additions to the Noēsis website, I continue to be awed by the originality of content. These ideas are every bit as sober as they are hopeful, and the diversity of subject matter is staggering. When Tony uses the word "sustainability," a term so often used today in our efforts to deal with climate change or pandemic, he's using that word in the context that most are not comfortable with. Tony means "human sustainability." Tony asks us to take a chance on clarity and ownership. He asks humans to either seek greatness, or to perish.

I repeat my above refrain. Tony doesn't describe a flawed humanity. He describes an injured and helpless modern human being, as fragile and as rare as it is confused. His arrival points are uniformly empathic. This book is not about buzzwords or trendy prescriptions for what ails humanity; rather, it is a

Big Picture analysis of why change must come. And what could be more important than that?

This book is must-read; it truly is. It's a clarion call for the salvation of humankind. Why is that needed? Tony has a direct response. Perpetual conflict is neither natural nor survivable. Tony makes us comfortable with uncomfortable subjects—a rare gift. But what can be done about these many concerns? Is there a prescription for human survival. Yes, there is....and it is in the book. Enjoy!!

David S. Jonas
Partner, FH+H Law Firm
LtCol, USMC (Ret.)
Senior Executive Service (Ret.)

Table of Contents

Introduction

The dinosaurs knew something was wrong. They were not just mindless predators or lumbering herbivores. Their instincts on this day swung over to non-specific fear. They couldn't see the approaching danger, but they felt it in every cell of their bodies. On this day, something was very wrong.

Every organism, whether primitive or modern, has well-honed threat responses. In their final hours, 66 million years ago, these magnificent creatures were riddled with anxiety. Their extinction event began as a vague beacon in the night sky. It was barely as bright as Neptune, and then Venus. A sentient observer may have called it beautiful. But on this day, weeks later, its brightness rivaled the Moon's.

This shapeless rock, this haunted traveler, arrived at a speed of 35,000 miles per hour. In the moments before impact, the asteroid approached the Yucatan Peninsula at a 60-degree northwest angle. When it slammed into the upper atmosphere, it compressed the air in front of it with biblical violence, ten million times the energy of a conventional nuclear weapon. And that was prior to touchdown.

We're just getting started here.

The dinosaurs had a 200,000,000-year run. They were not aliens. They were earthlings, utterly distinct from us, but still earthlings. The phrase *reptile brain* refers to just one of our many genetic inheritances. Neuro-anthropologists refer to something called "Triune Brain Theory." It's controversial, but it's a simple way of looking at how humans unconsciously respond to stimuli. This so-called reptilian brain includes the basal ganglia and brain stem. The theory divides the brain into three parts, with the oldest being the reptile brain complex.

The triune model locates "affect and emotion" in one of two places. It's either in the reptilian complex or the limbic system. Our ancient reptile brain senses threats and pumps the hormone epinephrine (also known as adrenaline) into our bloodstream. Our hearts beat faster, our breathing accelerates. Our senses sharpen. We become hyper-vigilant and reactive.

It's our reptile brain that reacts to a stranger's approach on a deserted nighttime street. The reptile brain keeps us from lingering in poorly lit alleys. The reptile brain might be called "The Sixth Sense," or a term even more familiar than that: instinct. Instinct, by its very nature, cannot think—quite the opposite. Instinct shuts down the thinking brain to do its job. "Instinct" and "Thought" cannot coexist in any given moment. This is an encapsulation of the modern human dilemma.

And therein hangs our tale.

The guaranteed and cat-quick overreaction to any given threat was vital to evolving humans to either vanquish the threat or to run from it. What might we expect to happen when the ancient dangers that our evolving ancestors once looked for are lessened and then one day cease to exist? With the ancient threats to evolving humans now missing from the modern human experience, did the ancient compulsion to overreact also lessen?

How could it? What cannot think, cannot alter itself.

These wrecking-ball compulsions are fully preserved in human DNA. Innumerable impulses and compulsions that were not designed to think never noticed the absence of their enemies. These ancient compulsions have fully reinvented, but worse than that, these agents of un-bargaining brutality still alive inside of us are making the rules for a degrading humanity. The driver of this jalopy called modern humankind is a battering ram. Its shiny buildings, its fast cars, its Prada footwear, its Armani suits—these provide the illusion of civility.

A deluded individual will seek remedy in a hospital, but a deluded modern humanity walks among the tombstones of its copy-cat failures. It sees an ever-deepening grave, but it can't read the most recent tombstone—its own. It can't view itself. I wish there were someone to blame. There isn't.

With the phenomenon of serial misinterpretation so rampant in the modern day, we hapless and helpless humans are on a bullet train to our own extinction event—but not while I breathe. Noēsis is here to help. I formed the non-profit called Noēsis in 2018 when I noticed something that we cannot expect to notice itself. I beg forgiveness in advance for what must now be said.

Not one of the world's accepted schools of philosophy is actionable. Roll that around in your head and keep the movie running. If it hasn't dawned on you, let me help it along. The world's schools of philosophy have suddenly combined to be a quantifier of human lip service and little more.

Noēsis is an actionable vehicle for human sustainability. We're here to identify and expose innumerable evolutionary mechanisms that cause modern humans to misinterpret everything they see. We're here to enable human re-integration with baser human compulsions that we sent to the basement. They still linger there, unchained. They want out. We will bargain with them, on their terms, not they with us.

Consider your own anxiety. If it's anything like what polls suggest, your level of anxiety is higher today than a few years ago. All humans on Earth are increasingly worried about our collective future. We feel the growing weight of it, the idea that this thing we humans do is on dead empty, but we can't see it. Clinicians call this non-specific or free-floating anxiety. To understand it better, consider a haunted house. When things go bump in the night, we're uniquely terrified because we can't see the multiple ghosts that are all around us. If we could see them, we could plan. How do we plan for what we can't see? Today's brand of nonspecific anxiety haunts all of humanity.

But again, we're just getting started. That 10-mile-wide asteroid is about to touch down. Wait for it. For now, consider this: a meta-analysis of almost thirty studies involving some 80,000 youth worldwide concluded that people under the age of 30 had the highest increase in anxiety post-pandemic. The prevalence of depression and anxiety symptoms during COVID-19 have essentially doubled, compared with pre-pandemic estimates.

What's happening to our young people? In response, I'll risk that the pandemic provided our children with their first hint that the sandcastle promises we made to them as well-meaning parents may run afoul of an indifferent universe. A beautiful song described it best:

"This is the end of the innocence."

Don Henley's song is more than a coming-of-age metaphor. Henley is describing human paralysis in the face of a gathering storm. Our children know that there's more going on than we tell them. A silent child is an intensely observant child. As loving parents, our hearts are broken to see our children harmed in any way. But our children must one day leave our loving arms. They will encounter pain that we will not be able to take back.

All of this would be troubling enough if it were merely a transitory increase in collective human anxiety. Instead, anxiety levels continue to rise. Are elevated rates of anxiety and depression anomalies, or are they connected to a bigger concern that isn't going away? Are we worried because of a pandemic, because of inflation, or because of global recession? Or are we feeling the same unshakable anxiety that afflicted the dinosaurs during their last moments on Earth?

If so, what are we prepared to do? Will we remain casual observers of yet another slow-motion trainwreck, inactive witnesses to an approaching maelstrom—just like the dinosaurs? Dino never had a choice, let alone a chance. Modern humanity has both.

Let's return to that asteroid.

At anywhere from 10-20 miles wide, it was not particularly large by cosmic standards. The asteroid is known to have broken off from the Oort Cloud that orbits the outer solar system, millions of years prior to its arrival. Now, most asteroids disintegrate in the Earth's atmosphere, so great are the forces of friction upon them—but not an asteroid this big. The asteroid impacted the Earth only three seconds after entering the atmosphere. It stopped dead on impact, and its kinetic force (mass x velocity or Kgm^2/s) then converted to heat energy. The water in the Gulf of Mexico instantly vaporized, and hundreds of trillions of grains of sand were super-heated into glass. Debris from the explosion blew up into the sky beyond our atmosphere and into space. It rained glass and sulfuric acid all over the world within 48 hours.

Over the course of only seconds, a crater 20 kilometers deep, and 110 kilometers wide that never existed, suddenly did. And that was the good news.

Due to the low approach angle, this cosmic invader plowed its way straight up to what is now Washington State. About 1500 miles east of that, thousands of dinosaurs congregated in what is now North Dakota. Their 200,000,000-year run on Earth was undisturbed before that. Evidence shows that dinosaurs were nearing the end of their reign by way of multiple natural phenomena.

On this day, perhaps Mother Nature decided to speed things up a bit. In any event, these dinosaurs that gathered in North Dakota could never imagine the evolutionary pink slip that was fast approaching. These were largely herbivores, minding their business, unaware of what was now barreling toward them from the South.

Only moments earlier, the Earth had been rung like a bell, but the asteroid's impact was not perpendicular. How could we know such a thing? Elementary! The leftover crater is vaguely egg-shaped. The asteroid's north/northwest vector is easily inferred. That alone — its low-angle trajectory — unleashed an ejecta cloud that looked from outer space like a 10,000 mph. wave of brown, super-heated mud. It engulfed the entire North American continent within 16 minutes.

The dinosaurs that gathered by the thousands on that day did not suffer. They were hit by a cosmic freight train, in the form of a two-mile-high wall of an advancing ejecta cloud. Fossil records show dinosaur skeletons so completely intact that these animals may as well have been freeze-dried where they stood. When the ejecta cloud lessened in what is now Canada, a residual wall of super-heated air continued its northern vector. Small animals found shelter. That bought them some relief, but it wouldn't last. Larger animals were helpless. They ran for their lives, never knowing what they were running from. Farther north, herds of Triceratops experienced an immediate temperature increase from 90 degrees Fahrenheit to 250 degrees. A lazy stroll through a peaceful valley was suddenly an infinite oven of flash fires and boulders falling from the sky. A typical dinosaur's lifespan was 30-50 years, but this 1000-mph wall of 250-degree super-heated air meant that their lifespans were now measured in seconds.

A magnitude 12 earthquake was next. The strongest earthquake ever measured on Earth is 9.4, the Valhalla earthquake. Anything above magnitude 10 requires an external force, and the asteroid provided it. The quake continued uninterrupted for six months. Such an earthquake today would end all life as well as destroy every man-made object on Earth.

The avian dinosaurs had an advantage—or so they thought. They went airborne and stayed there. Inconveniently, the electro-magnetic pulse produced by the impactor arrived and ignited a lightning storm that covered half the Earth. In no time, the avian dinosaurs' wings were pelted from above by fiery gobs of sulfuric acid that rained down on them. They were knocked out of the sky one by one, like carnival ducks at an arcade.

Virtually every coastline on Earth was impacted by mile-high mega-tsunamis within 24-36 hours. Rocks that originated at the site of impact are found today as far away as New Zealand.

• • •

The Earth was different then. If a modern human emerged from a time machine into the atmosphere of 66 million years ago, he or she would not survive the science that ruled over these ancient realities. Filled with plant life, the Earth was a constant machine of photosynthesis. Earth's atmosphere was far too oxygen-rich for the lungs of modern humans. That reality would change overnight, in cosmic terms — a few weeks, a few months, and for any living thing over five inches tall, a few decades.

Debris from the asteroid blanketed the Earth in a "nuclear winter" that blotted out the sun for a year, as Earth's temperatures plummeted. The dominoes fell in precise order from there. The Laws of Physics do not bargain. With photosynthesis impossible, most plant life was killed off. This in turn starved the herbivores. When they died off, the carnivores then lost their primary food source. In mere decades, 85% of the species on our planet ceased to exist.

The Late Cretaceous Period ended in one day. That's a fair statement.

Or is it?

You see, the asteroid itself did not kill the dinosaurs. The asteroid committed the dinosaurs to their demise, but it was the effect of the asteroid over time that produced the finality of extinction, and that would require decades. These two vastly different concepts — commitment and finality — are the phenomena that a degrading modern humanity must now view.

And that's precisely why I created Noēsis. I've placed this humanitarian group at the forefront of these somber but vital processes. We'll need to earn it. Count on it. Once humans understand the meaning of irrevocable commitment to extinction, we cannot allow it to escape our gaze so long as we walk this Earth.

• • •

I don't ask that you take time to read this book; I ask that you make time to read it.

I'm not in the business of describing extinction events. I'm in the business of preventing them. But we're going to briefly move on from human extinction. Let's talk about a thief in the night: Commitment. Commitment is the mathematical point past which a heavily deluded modern humanity will never return, no matter how many rationalizations intrude. Should the human species ever reach the point of commitment, it would only be known in retrospect. Try to imagine the collective horror of an innocent and helpless humanity when a cosmic masterpiece called *Homo sapiens* learns that recess is over—the day that ignorance is no longer bliss.

The dinosaurs felt their approaching reality in that moment. Humanity will learn of the commitment from which it cannot recover long before the Earth falls silent. Should humanity achieve commitment, call it "Cosmic Checkmate." If modern humans invite that day by either our actions or our inactions, our children and the unborn will know existential hell on Earth. I ask again, what are we prepared to do? If your answer is "nothing," you may place this book down. You have your reasons, and I respect them. For the rest of you, let's know what is now knowable, to prevent what is preventable.

What about Mankind? Have we committed to our demise, and if so, how can we be sure? Frankly, we can't. That right there should chill all of us to the bone...only it doesn't. One thing is undeniably certain: should humans achieve commitment, we can tell ourselves any lie we want, the idea that we'll figure it out, the idea that this could never happen to us.

We will not figure out commitment any more than we'll calculate the precise value of pi. Commitment means that we've clocked out. Leaving the building will be a hell so unimaginable for our children and our descendants, and I won't sit still for another moment or for any more rationalizations.

These pages will demonstrate the mathematical certainty of an approaching commitment that, for the first time, we can understand. For thousands of years our species has grown, evolved, expanded, and learned. But at what cost? Let's look at the scoreboard:

With the deadly precision of a radium clock, over 40 stationary civilizations of the last 10,000 years have come and gone. History books describe a myriad of variable situations. Situations sell! However, the "situations" that our history books cite are now known to have been more correctly symptoms. Human interaction never once failed by situational phenomena. We fail by a socio-mathematical constant, not by its situational variables. For the greater good, I risk my popularity with historians by claiming with certainty what has never been claimed before. Human history books sell because of their situational descriptions. Situational analysis of human failure is as useless as a graverobber in a crematorium. The core reasons for machine-like human collapse, while uncomfortable, will resonate even among the most stubborn historians, in the privacy of their minds. There will be no concession from these historians. I require none. The brutal nature of socio-mathematics will rule these pages.

Let's get on with it, and not a moment too soon.

What philosophers have dubbed "The Human Condition" is an abandoned mansion. What we think we know about ourselves very suddenly reveals a clueless and deluded humanity. Call this book Human Condition Redux. For the first time in history, we can understand who we truly are. It was the bad news that kept our evolving ancestors alive, not the good news. Let's strap ourselves in and test that theory.

Use this below QR code or link to watch a slow-motion cosmic trainwreck, a visual depiction of what happened on that awful day, 66 million years ago.

https://www.youtube.com/watch?v=vq3nWnTkFbk

Part 1

Understanding the Human Injury

Chapter One
Noēsis

The more you know about Noēsis, the better for us. But for now, Noēsis is largely a stranger. And we humans are suspicious of strangers! So, take all the time you need, and ask questions. Let me get you started. What is Noēsis? Why must Noēsis exist?

Noēsis is a non-profit organization dedicated to helping the human species achieve sustainability. Why is human sustainability a concern? It goes like this: a helpless and exhausted humanity has suddenly arrived at a place that feels far less promising than we were taught to expect. The human species, all eight billion of us together, senses a gathering storm. Noēsis was formed to help humanity achieve sustainability by exposing the evolutionary motivators that combine to influence all modern behavior. Ancient messages, the residue of 2 million years of evolution, haunt the human mind.

The only way to get ourselves under control is to determine how we lost control. We have a key that was never available–the evolutionary data that will allow self-discovery for the entire human species.

Our mission is to identify and describe the machine-like action of evolutionary forces inside all of us, against which we are largely powerless. Bringing them into view will open a window to human sustainability.

Humanity's attempt to become a sustainable species is further challenged by a random crossover event that occurred in our evolutionary timeline. Due to this crossover event (herein dubbed The Human Injury), humans are acting in a technically unnatural fashion. Unnatural behavior by itself will select any living thing for extinction. Humans are committed to their eventual demise until this phenomenon is completely understood, and until actionable methodology is deployed to alter it.

All eight billion humans on Earth have 50,000 generations of collaborative, fraternal and affiliative behavior woven into their DNA. The crossover point, from sustainable behavior to unsustainable behavior, occurred when collaborative human groups reached all areas of the globe, and settled into stationary populations. Collaboration was discouraged and replaced by Acquisition.

Self-interest and collaboration repel. Acquisitive humans were now acting unnaturally. The collapse of 50 human empires, by this crossover event, was assured.

The Human Injury is a deal-killer. What could that possibly mean? It means everything. Humanity will understand it in detail or be consumed by it yet again.

We can all agree that it's not natural for humans to breathe in water. If any human being breathes in water, they will die straightaway. This ridiculously simple analogy describes a much broader point. For the last 15-20,000 years, human beings have inarguably acted in ways that are unnatural to us. You must demand that I prove it, and so I shall. If what we hold to be true is nonsense, Noēsis has no future. Don't count on it, friend! Noēsis — or an entity like Noēsis — must now exist if humanity expects to arrive at a sustainable future.

I issue the following instructions for the faint of heart. I'm about to make you uncomfortable. I have no choice. Dawn comes after dark, not the other way around.

Anthropological computer models assign a probability of 1 to 2% for humans of the 21st century to arrive at a sustainable future if our current inertia persists. Here's where I risk my popularity yet again for a future reward. We will perish from this earth if we don't bring ourselves under control. Moreover, these concerns will be on top of us sooner than we think. If the brutal intrusion of COVID surprised you, have the courage to look around. Behold the totality of human degradation and extrapolate from there. Whatever drug you're on, if it allows you to see to the other side of the current human inertia, perhaps we should put it in the water supply, like fluoride because I'm not seeing it.

You want to run away now, right? It's that very phenomenon — pain avoidance and pleasure seeking — that allows that probability to grow. Avoidance sentences us to an unknown fate. Choose clarity!

At Noēsis, we hear this a lot: "Humanity can't unite because we live in separate countries." That's complete nonsense. We all work for different companies, right? We all went to different schools. We all grew up in separate homes. When did that ever stop us from getting together at the end of the day to enjoy one another? We get together as one because we ARE one.

Can humans unite politically? Nope. Can we unite religiously? No, we can't. The multiple gods that divide modern humans will kill us deader than Julius Caesar, perhaps within a century. How about culturally? Again, no. We're not there yet, and we're not even close.

But what about species preservation? Let's agree on the following:

- Political, religious, and cultural unity is not remotely possible.
- The preservation of the human species is a visible cause under which all humans can unite right now.
- Recognize this emerging fourth banner, and nothing more. Politics, no! Religion, no! Cultural, no! The preservation of the human species. Yes!

• • •

As children, we were all afraid of the dark. Being afraid of the dark makes us human. The observation serves us well here. There's a vague darkness settling in over our entire species. We all sense a gathering storm. Our denial, our false bravado, serves us poorly.

This book intends to move human anxiety out of our peripheral vision and into the spotlight. If we bring human anxiety into view, we rob it of its ability to barge in on all of humanity. Let's hear ourselves say what we never could. All eight billion humans share a common paradox. Let's shout it out. Are we ready?

We don't know how we got here, and we don't know how we're leaving!

What sentient being would not fear that reality alone, let alone the advancing degradation all around us? If you've got a heart beating in your chest, you sense freefall.

NOĒSIS

THE HUMAN INJURY

COLLABORATION
YIELDS TO
ACQUISITION

- ILLUSION OF OWNERSHIP
- UNNATURAL HIERARCHIES
- GREATER/LESSER BEINGS
- DNA-BASED INSULT
- UNRESOLVABLE CONFLICT

= Ø

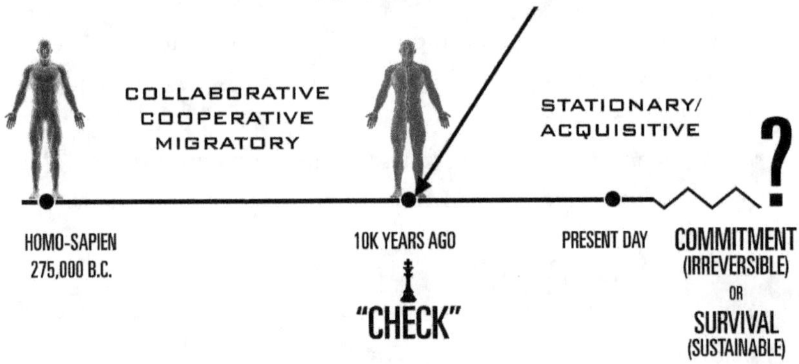

CONCLUSIONS

- SELF-INTEREST IS UNNATURAL; HUMANS HAVE DEVOLVED
- UNSUSTAINABLE/UNSURVIVABLE
- REDIRECT OR FAIL

SCAN THE QR CODE FOR A DETAILED EXPLANATION OF THIS ILLUSTRATION.

NOĒSIS

POST-INJURY COLLAPSE

"MATHEMATICS OF SOCIETAL BREAKDOWN"

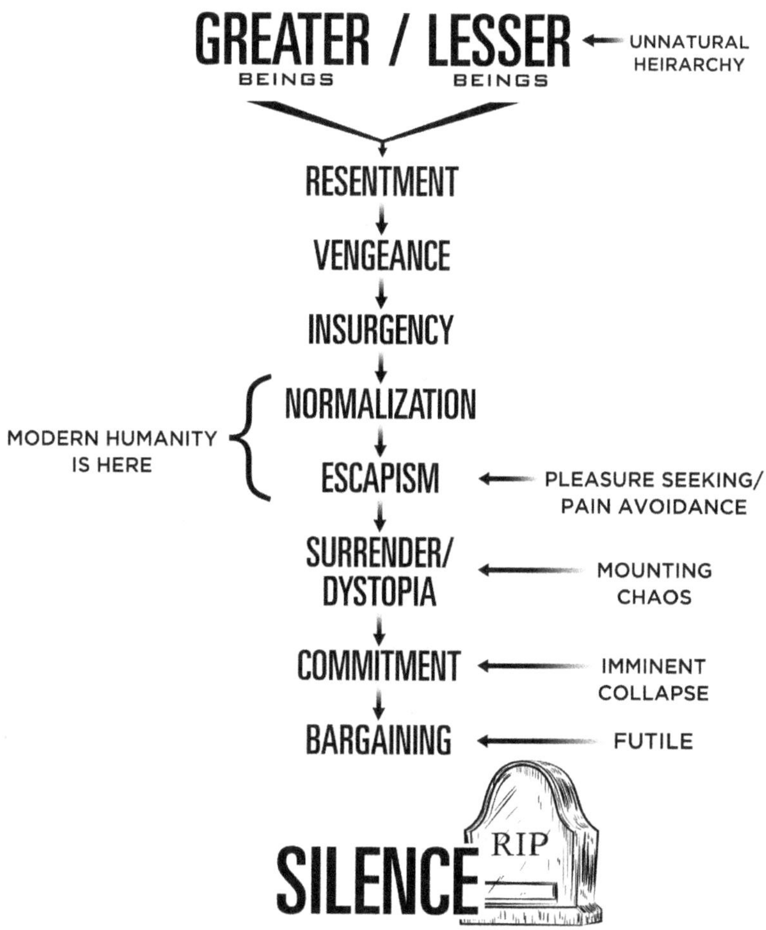

SCAN THE QR CODE FOR A DETAILED EXPLANATION OF THIS ILLUSTRATION.

A person with a parachute is in blissful freefall. But what happens when they pull the ripcord and determine with horror that the parachute doesn't exist? The bliss experienced only seconds ago is transformed instantly into a grim reality. Similarly, freefall for all humans is no longer conjecture. The freefall we all feel doesn't hurt. The sudden stop at the bottom will. Noēsis is here to redirect that possibility because the sum of human fear has suddenly paralyzed all humans.

Here's what paralysis looks like. We don't know what to do or where to start. That produces the staggering indifference we now see. "Indifference" and "Escapism" don't fool me in any way. False bravado, by any other name. Here's what's really going on: we're frightened. That's a common enemy that all humans now share. Human history proves that humans unite over common enemies. Can we unite over the specter of our mortality — is that not enough? We're about to find out. Paralysis goes nowhere! Paralysis is tantamount to surrender. I will not surrender—ever. Will you?

THE HUMAN HEART IS BROKEN

The collective human mind is gravely injured. The collective human heart is broken. It broke long ago, when we gathered into stationary cultures and were asked to choose sides against our brothers and sisters. Perpetual conflict – the serial destruction of our own – immediately runs afoul of an inarguable truth. No human on earth has a natural enemy.

We are fraternal and helpful creatures at our core. An event in our evolutionary timelines nudged us slowly away from that. A disturbance to the fraternal inertia of evolving humans had no choice but to eventually arrive at crisis. And here we are — in crisis. Our history books tell us that human failure is situational. Situations sell history books. Their utility ends there. The current human dilemma is a constant mathematical arrival point.

Would we punish an injured animal? No. Would we scold a lost child? Of course not. When we view our entire species as injured and lost, a clueless humanity might suddenly like itself enough to preserve itself. Can we get there? If we don't, get ready to blow taps for the most remarkable species to ever exist.

We won't stop human freefall, but we can install that parachute of education. What's our choice? The sudden stop at the bottom? Never.

Human sustainability is not a zero-sum game. It's about maximizing probability. Humanity is on the clock, and there isn't anyone reading this who doesn't plainly know it.

AS AN ASIDE!

I should explain something about this book straightaway. As the founder and president of Noēsis, I planned and then created an actionable vehicle for human sustainability. This nonprofit entity was built for the purpose of problem-solving, to allow a spiraling humanity to find a sustainable future. For anyone who feels that we have enough human sustainability groups, you won't feel that way once we're finished here.

This book is an exercise in counterintuitive thought. What modern humans call "intuitive" has now decayed into a bleak and barren landscape. We're a failed culture —locally, statewide, countrywide, and globally. The sum of human degradation is no less than astonishing. The brutality of that pronouncement will hit you every bit as hard as your inability to defeat it. If you seek paradigm-shifting thought, take that elsewhere. I call it common sense on steroids. I have a persistent and annoying habit of making sense, and I know my business.

Earlier, I asked permission to make you uncomfortable; now I ask you to be inconvenienced. Truth is always extinguished, never congratulated. The truth you see below is immune to scrutiny. I will entertain arguments to the contrary, and I will concede to material fact, as required by the principles of critical thinking.

THE CASE FOR SUSTAINABILITY

There are four major concerns at the forefront of human sustainability efforts. Climate Change, Pandemic, Nuclear Non-Proliferation and Disruptive Technologies. These sincere and caring missions suggest that we cherish our existence, even as the end of our existence is visible on the horizon. We deserve to survive. But you're thinking...

> ***"Deserve's got nuthin' to do with it."***
> —William Munny, *Unforgiven*

I'm aware that the word deserve will attract cynics. They'll default to provable nonsense—that we humans have been bad boys and girls. Is that so? Anyone who labors under the cognitive distortion that humans are a scourge should check their misanthropic defaults at the door, and we will view the modern misanthrope forevermore with empathy. But first, let's shed our biases and become uncomfortable.

- Should human beings execute plans to curb ***climate change***, it will move the needle of human sustainability in no way whatsoever. Please accept my deepest apologies.
- Should the awful phenomenon of pandemic (lately dubbed biosecurity) be forever contained, alas, the current human trajectory is not survivable. We will perish by our own hand. The word "pandemic" will not appear in the ensuing autopsy.
- Should humanity achieve the miracle of disarmament, where every nuclear weapon is disassembled, we assume that our chances of surviving to Star-Trek-like sustainability will skyrocket. With the heaviest of hearts, I cannot concur. If disarmament occurs on a Sunday, we'll be re-armed by Monday. Evolutionary paranoia will not submit itself to be disarmed.
- The most recent concern for humans, one broadly called disruptive technologies, cannot be diminished in importance. Should we control it, we will do so just in time for our burial in the cosmic graveyard of past cultures that never deduced their only true nemesis.

If the above four major concerns were to be controlled, a sufficiently deluded humanity would throw a global party. The party will feature all the vigor of a pep rally where we sing fight songs to ready us for the big game called sustainability, the sustainability that we all assume we've achieved. That pep rally, should it ever occur, will forget that 50% of pep rallies are in vain. Tomorrow, the victory will not go to the better pep rally. It will go to the better team. The team that modern humanity plays every single day is undefeated, unscored-upon, and fully invisible. The score in tomorrow's game is precisely knowable. Fifty to nothing. 50-0. [Fifty represents the number of stationary human cultures of the last ten thousand years that played this very same opponent. They were crushed by what they never saw. They never had a chance].

With the "Big Four" controlled, we will have spent trillions on these efforts. When these four concerns, the ones that occupy the forefront of human anxiety in modern times are brought under control, we will likely have committed us to our demise by that time. Our sustainability bliss will be rudely awakened by what will never admit its existence, let alone allow its detection.

NOĒSIS

— HUMAN SUSTAINABILITY INVESTMENT —

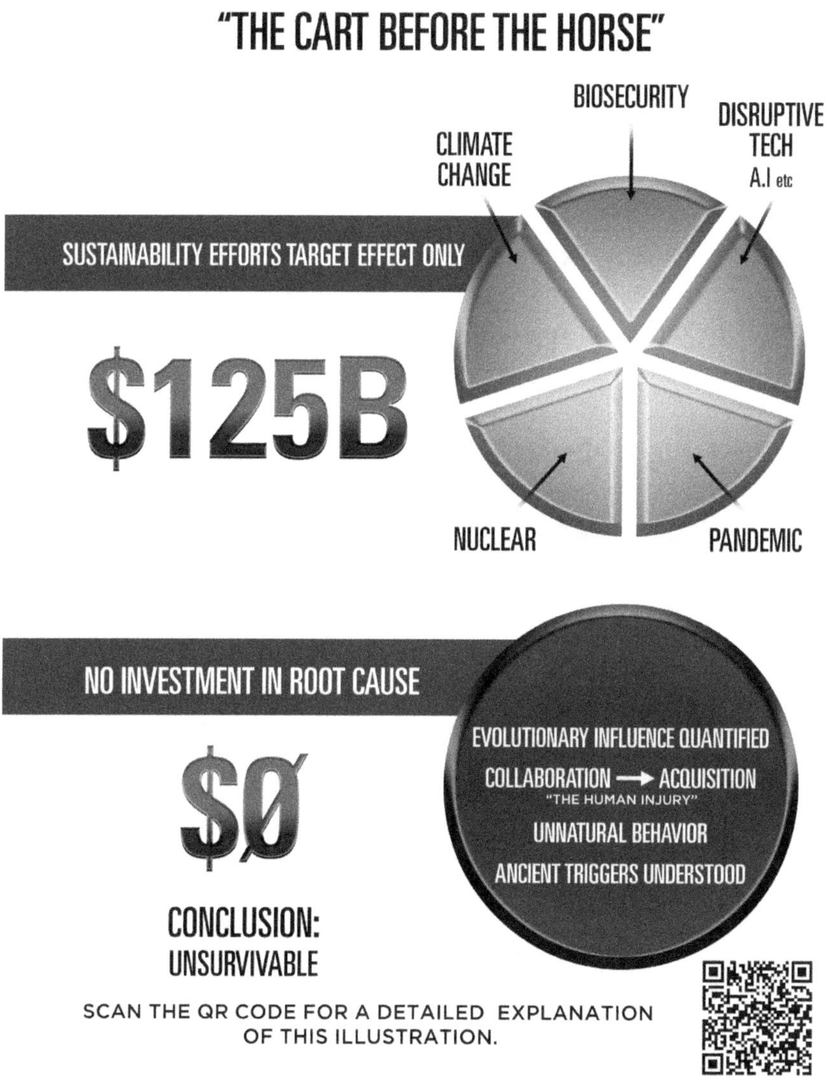

SCAN THE QR CODE FOR A DETAILED EXPLANATION OF THIS ILLUSTRATION.

WHAT'S THE BIG MYSTERY HERE?

You see, folks, that's just the thing right there. There is no mystery. By its nature, any mystery features an absence of one or more clues, ones that must be discovered or our mystery cannot be solved. The clues that you seek are not merely accessible; rather, they hang off modern humans like dead skin. For anyone who feels vaguely irritated, wondering if this is a game, your irritation is further proof that hiding in plain sight is the only correct strategy for a serial killer. Indeed, serial killers are known to revisit the scene of their crimes and stand there in the crowd, admiring their latest calamity, as clueless investigators look right past them.

The analogy is quite precise. Modern humanity routinely assigns complexity to exquisite simplicity, any tool it needs to avoid seeing it. Not one of the four deterrents listed above, ones that humans feverishly attempt to control in the present day, has ever been found on the tombstone of a dead human culture.

By now, you're bursting at the seams. Would I disrespect any reader in this fashion? I certainly would not. Still, I ask that you endure as I set up the last bowling pin. Many will recall a famous psychological experiment that demonstrated a phenomenon called *inattentional blindness.* That term is not to be confused with *willful blindness.* In the latter, humans knowingly look the other way as a moral or ethical concern accelerates.

The phenomenon is also known as *inattentive blindness*. In this test, viewers were asked to focus their attention on one task. They were asked to watch a short video of a group of 9 or 10 people, aimlessly milling around. Three of the people in the group were wearing white shirts. The test viewers were asked to count the number of times that the three people dressed in white passed a basketball to one another. With these simple instructions understood, the video began. Each viewer focused on counting the number of times the ball was passed. There was no pressure; there was no penalty. The viewers simply needed to focus hard, due to multiple passes and the presence of several other people not dressed in white.

After watching this short video, the test viewers reported the number of passes they observed. Most counted the same amount (15). The answer was correct. What no one noticed during this video was that a person with a gorilla suit walked slowly through the frame. The person in the gorilla suit walked slowly and deliberately and even stopped in the middle of the crowd to take a bow to the viewers—not that it mattered. The majority never noticed the gorilla.

What did this innocuous test prove? We well-meaning human beings are oblivious to what goes on around us if we are on task. Look at this link, to view the experimental video, or take my word for it. Your choice.

https://www.youtube.com/watch?v=vJG698U2Mvo

Willful blindness, as noted, is fully purposeful, but this book isn't about blame. Inattentive blindness has no one to blame. With modern humans "on task," we cannot see the calamities all around us, and that cannot be blamed on anyone. The forceful rejection of blame occurs billions of times daily. Neither blame nor judgment has utility of any kind. To be clear, an innocent, confused, broken-hearted and paralyzed modern humanity has a bad case of itself, in the form of acquisitive inattentive blindness. The phenomenon has produced a mountain of normalized calamities. A mathematically guaranteed avalanche of normalized calamities is just ahead. A self-loathing humanity will despise its own reflection yet again, having in fact committed no crime whatsoever.

The phenomenon of normalization will be lessened if our species intends to survive itself. For now, you're still waiting for me to drop the bomb you still can't see, and you're fresh out of patience. I brought you here purposefully. Call it tough love. Look here:

Earth has hosted innumerable stationary human cultures over the last 10,000 years. Many of these cultures were unaware of the others. Some were small, some were medium-sized, and at least forty of them reached "empire" status. These include Mesopotamian, Ottoman, Babylonian, Egyptian, Greek, Holy Roman, Byzantine, Inca, Mongolian, British Colonial, and others. These cultures collectively averaged 300 years each. Each was dead on its feet long before that, disassembled and destroyed in the same precise fashion. I stress the word *precise*.

None of these cultures experienced climate change. Pandemic visited all of them but wiped out none of them. None had nuclear concerns. None had rogue computers. The tombstones of all major human empires on Earth had only two words on each. No other words were needed. No other words are needed today, as modern humanity stands squarely at the door to oblivion, being handed its hat by a catastrophic phenomenon it can't see. I thank all readers for following along. The only two words on all fifty tombstones were these:

Perpetual Conflict

For those who view this above revelation as anticlimactic or myopic, we need to talk. You're still counting the number of passes in that video made by white shirts. You've yet to glimpse that smug and entitled gorilla that walks through that video every time. You didn't deduce that the experiment never cared about the number of passes. It cared only about how many people noticed the gorilla. You can be certain that I noticed the gorilla. I wrote a book about him. It's due to hit the shelves the moment we wake up. Whoever does not see perpetual conflict among humans as being the only demonstrable killer of human interaction is blissfully "on task." "Perpetual Conflict" walked through the video of fifty stationary empires on Earth. We never noticed it; we were too busy feeding it.

May I please drop another bomb of clarity? Brace yourself.

Modern humanity is a failed culture. If any reader is aghast over the weight of that declaration, I encourage your indignation. I invite any counter-argument that would defeat the observation. When none appears, it means that I just told you in one sentence what you could not tell yourself just ten seconds ago.

Is this a new low for humanity? Many assume the answer to be a resounding, "yes.' Again, I will explain to you what you have not considered. This is not a new low. It's the LATEST new low. Remember the scoreboard? Fifty to zero. The constant is "stationary culture." It's ground zero for human failure, and I know why.

Look at that gorilla called, "perpetual conflict." Don't ever look away.

• • •

We specialize in macro-humanism here at Noēsis — the study of large groups, up to and including the entire human race. As a collective species of like and equal beings, currently positioned at one another's throats, call this the mother of paradox. What sustains us, what puts food on our tables, what educates our children, is also our only known executioner. Under these circumstances, we will not survive ourselves. Any disagreement, any idea that the modern inertia of humankind will survive its own madness, will require a defendable argument. Here are a few examples of our present-day, sandcastle rationalizations that will no longer be tolerated:

- *"We'll be just fine!!"* Your protest is noted. It is not an argument. Use it as an excuse to allow further human degradation, at great risk.
- *"It's only at the precipice that we find the will to change."* This is no movie. This isn't a game. We're on the precipice right now. Beyond it lies the abyss.
- *"We'll figure it out."* No, we won't figure out commitment. Fifty dead empires said the same thing. The latest culture, the one reading this book, is the 51st empire. We're failing again, by the same models of conflict into perpetuity that wiped out all others.

That gorilla is the only thing I see. I invite the world to try it—and to live.

• • •

Conflict among humans is not natural, other than defensively. It required just a few short paragraphs to describe the only demonstrable reason that humans fail. The other four — climate change, pandemic, nuclear, and disruptive technologies — are secondary. And yet, here we are again. We notice the cough, the sneezing, the sniffles, and the runny nose, but never the flu. The constant flu that destroys human interaction is perpetual conflict. There are no exceptions of any kind. What would good news look like in this assessment? It's this. The components of socio-mathematical human failure are now eminently knowable. Humans are out of excuses, out of time, and out of clichés.

"All these moments will be lost in time, like tears in rain."
—Roy Batty, *"Blade Runner"*

• • •

Just when you thought it was safe to go back in the water, here comes yet another torpedo of clarity, straight to the bow of a deluded humanity. Modern human beings act unnaturally. Humans have been practicing models of unnatural hierarchy for 200 generations (200 generations x 50 years each = 10,000 years). Prior to that, we practiced 100,000 generations of natural hierarchy. Those residual evolutionary longings are our deepest ones. They burn every bit as brightly inside of us as they ever did. They call us now to resurrect them, to save them.

By contrast, unnatural hierarchy is mathematically doomed at its inception. Its cookie-cutter failures call to us as well, to rise from the ashes of our failures, only to re-ignite them. When we do, the starting gun of unnatural hierarchy kills our horse. It staggers out, mortally wounded, and never recovers. Our new cultures look promptly like the old ones, and fast. Our history books write it all down, convinced that these clock-like failures are "situational." That distortion – that human failure is situational— must be retired forever. Historical "situations" are variables. All past stationary human cultures were dead on their feet. The constant socio-mathematics that guarantee such a bold assertion were ignited by a fully viewable evolutionary event. We can see back in time, not only to the Big Bang that likely created our universe, but also to the evolutionary Big Bang that pushes a helpless modern humanity to an ending that never needs to happen.

This lone evolutionary crossover phenomenon by itself derailed a successful, evolved, dignified, sustainable, and loving masterpiece of creation called *Homo sapiens*. These longings were forged by fifty thousand generations of migratory, cooperative, and collaborative perfection. The intrusion of unnatural hierarchies ignited a DNA-based insult. That crossover period forced pristine, collaborative human DNA to slam on its brakes, throw it into reverse without a clutch and to deal with its new normal. Collaborative human DNA was given a pop quiz that it never studied for, administered by an indifferent universe. The clocklike failures would go on to be indistinguishable from one another. A mathematical constant guarantees the precision we can now observe. We can shorten it up. 50-0. Or we can work it—and live.

Two hundred generations removed from fifty thousand generations of global wandering, and with fifty major stationary cultures having already failed in precise fashion, a modern human species that deludes itself to be advanced has in fact devolved into what no longer recognizes itself at any level of its DNA. If the idea of a devolved modern humanity outrages you, again, we need to talk.

Consider a diseased liver. Once a healthy liver commits to an irreversible cirrhotic state, it breaks down with well-documented precision, eventually destroying its host. Human organs routinely devolve from their formerly healthy disposition. Just the same, the modern human being can see its prognosis clearly, will not view it, and therefore cannot escape it. What will happen to future humans is not in any question whatsoever, such is the brutality of a socio-mathematical breakdown. For the first time in human history, what WILL happen to an inattentively blind humanity no longer NEEDS to happen.

Are we uncomfortable yet?

The human species is devolving into hell on a rail. We are accelerating, in the very same way that water accelerates as it circles a drain. Any argument that views modern humanity in its current iteration as sustainable in any way is flatly obtuse. That's a mic drop, and a purposeful one. Someone must do this. If modern humanity will not view its prognosis, I'll sign the death certificate right now. Expedience is indicated.

For additional context, consider a popular character from modern culture. Let's ask, "The Wolf" how he views the human dilemma.

We all remember, *Pulp Fiction*, the Quentin Tarantino masterpiece. Its ensemble cast formed a who's who of walking metaphors. One of them, Winston Wolf, played by Harvey Keitel, was sent in to assess and clean up messes that were deemed uncleanable. In my mission to shake modern humanity out of its haze of indifference, I evoke the same urgency that Winston Wolf expressed. Our task, to achieve human sustainability, is as immense, and suddenly just as urgent. Time is a luxury humanity no longer enjoys. We need straight answers. Will you look or avert your eyes? The Wolf insists that we choose clarity, and would we please hurry it up!? The Wolf put it like this:

> "If you don't mind, **pretty please,** with sugar on top."
>
> —*The Wolf*, "Pulp Fiction

UNNATURAL BEHAVIOR

Before we continue, let us regroup. I ask again, are there any arguments that defeat the brutality of the above analysis? Please talk among yourselves and forward an alternative that defeats the stark nature of my observation. I must first issue a gentle and respectful warning. Like all readers, I know my business. Any argument you bring may collapse under an avalanche of facts. I have command of these facts. Theory puts me out of business, and it should. There is no complexity of any kind. Watch how easy this is.

"Unnatural behavior cannot persist in nature."

And this is complex?

Stationary human cultures automatically default to what is not casually unnatural, but rather, wholly unnatural, relative to the affiliative communion that humans shared during a migratory journey that required all hands on deck, at risk of catastrophic failure.

This Earth will shake unnatural behavior off its surface like a dog shakes off water. This thing that stationary humans now practice has always destroyed us. It's destroying us now; it will destroy us again. If the brutality of these descriptions rankles you, then what are you prepared to do? Will you throw this book down in disgust, or will you engage?

• • •

A 250,000-year linear masterpiece of human evolution produced a miracle that arrived at every corner of the Earth 10-15,000 years ago. The journey began for all humans in Central Africa in the African Congo Basin. That journey now links modern humans genetically by 99.9%.

Brothers and sisters are at endless war—like and equal beings, positioned at one another's throats, fighting battles that were never meant to be fought, waged against enemies that never existed. No human on Earth has a natural enemy.

My God, what happened? The Human Injury is what happened. Unnatural behavior, birthed by a random crossover event, from collaborative perfection to self-interested catastrophe, created the eight billion individual islands of self-interested, and thereby unsustainable humans we see today. It has never survived. It will not and cannot ever survive.

Deductively, therefore, stationary human cultures of the past enjoyed not one single second of sustainability. If these cultures were metaphorical gardens, their flowers died the day they were planted. This very same unnatural behavior, perpetual conflict in plain sight, is why modern humanity is inarguably dying.

Natural selection assigns no premium whatsoever to its creations. As modern humanity accelerates into an existential curve, we will neither be bargained with nor warned. In a geological second, with no fanfare of any kind, this planet will fall silent. In 20,000 A.D., our bones fully powdered by then, a grateful planet won't recall that we were here. With these deadly constants all around us again, what will happen to modern humans is not open to interpretation any more than 2 + 2 is open to equaling 5. Only a process-driven methodology, unfolding over decades of slow-drip education, using evolutionary data now available will allow self-discovery for a frightened and broken modern humanity to understand itself.

Should our species not partake of the educational meal available to us, it's all over but the paperwork.

Who Am I, You Ask?

I can feel pockets of indignation brewing in the minds of several readers. If your indignation could talk, what would it say? *Who is this man, and who is Noēsis to make statements of this outrageous nature?*

I'm a man who makes and defends arguments. I'm a man bound by the discipline of critical thinking. I'm a man of fearless nature, willing to shake a deluded and helpless modern humanity out of its malaise, for the most noble of reasons. As the river of escapist human behavior treks toward the ocean called surrender, that arrival point will not be considered in any way. It is my intention to build a consortium of global empathic thinkers. It is my intention to unite all humans on Earth under the banner of species' self-preservation. We need leaders, with answers. Noēsis will take the lead if you'll let us. We'll earn it. It starts with this book. Human degradation must stop NOW. The sandcastle rationalizations that I earlier mentioned have gathered here in this book. My message to them is succinct: you're fired!

Who am I? You can call me Tony, so long as you recognize perpetual conflict as being a mass murderer that makes Ted Bundy look like a piker. The astonishing degradation all around us, fully normalized, fully monetized by an emotionally exhausted modern humanity, has the human species staring down a shotgun barrel, should we not abandon our innumerable delusions. Humankind will either slow, arrest, and reverse advancing human degradation, or we will attend one last pep rally, to dance in the ruins of our inaction.

Am I saying that these glaring, socio-mathematical constants will fail once again, because of the perpetual conflict they invite? That statement is suddenly not a strong enough one. We have already failed. The helpless modern human being is paralyzed.

I say again, the stark nature of these descriptions is purposeful. I intend to shake the cage of a deluded humanity. These brutal descriptions are existential sledgehammers, but they have their place, especially when what's in front of us will not respond to coaxing. With many sledgehammers sufficiently deployed, turn toward the only logical question now:

Why is the only provable enemy of human sustainability — conflict into perpetuity — not listed as a deterrent to human sustainability? This book will solve that mystery. Like all mysteries before it, it was "The Butler" who did it. The Butler now has a name and a face: perpetual conflict, in plain sight.

I don't ask questions I can't answer. What failed system admits its failure? Why would our fully monetized model of global conflict run itself up the flagpole? Perpetual conflict watches now as a helpless and exhausted modern humanity creates three brushfires of conflict for every one brushfire that we stamp out. This is not the eternal flame we counted on. Ceaseless and mindless conflict is undefeated. We will redirect, or we will perish.

A surrendering humanity will view itself as children at a birthday party. We blow out a trick birthday candle of conflict, and we watch in wonder as it reignites. Our childlike wonder will cross over to adult horror, or we will be swept away by an indifferent universe that neither knows nor cares that we exist.

• • •

Evolving humans were successful, dignified, and loving beings, but the icy hand of natural selection had no medals waiting for us. In retrospect, therefore:

- Evolving humans never knew that the stationary cultures at the end of their migratory journey would default to unnatural models of acquisitive meritocracy.
- We deduced evolution in 1859, when Charles Darwin published, *On the Origin of Species.* We worked on every mystery without the first clue. A default to paranoia was the only arrival point for a miracle of creation that could not define itself.
- We never knew that resentment would one day intrude on our sustainable bliss, when stationary cultures commanded us to choose sides against our brothers and sisters.
- We never knew that the evolutionary crossover from successful collaboration to acquisitive frenzy would ignite a socio-mathematical constant of assured failure.
- We never knew that these assured failures would rise from the ashes and start the same failed play all over again.
- Pick out anything that we now know in the modern day. We didn't know it then. This innocent being, a "1 in 300,000,000,000 odds" miracle of creation, is the same being that we blame today for not guessing all the right answers. I weep for this depth of confusion; I do not condemn it.

OUR ANCESTORS COULDN'T DO THE MATH BECAUSE THEY DIDN'T HAVE THE MATH. EVOLUTIONARY DATA THAT MODERN HUMANITY NOW HAS IN GREAT ABUNDANCE WAS UNAVAILABLE TO OUR ANCIENT ANCESTORS. OUR TIMELINE ON THIS EARTH IS KNOWN IN DETAIL. WE'RE FRESH OUT OF EXCUSES, MY FRIEND, BUT WORSE, WE'RE FRESH OUT OF TIME. RECESS IS OVER. IF NOT, THE DEATH CERTIFICATE WE DON'T DESERVE IS ON THE HORIZON. WE WILL NOT CHEAT THE DATA, AND WE WILL NOT CHEAT AN INDIFFERENT UNIVERSE.

THE FLIGHT MANUAL

What happened to humanity to derail it? Something happened to us — an evolutionary crossover event that is now understood in detail. How could we have known it prior to 1859, when the flight manual for a clueless humanity was written by a man named Charles Darwin. For any Anthropologists now reading this, resist the urge to isolate minutiae, for the purpose of punishing it. Instead, recognize a huge promotion when you see it. This book will promote the work of socio-anthropologists all the way from important to central to human survival. The collective work of socio-anthropologists is no less than a road to human self-discovery. Who we are is now knowable in great detail. The evolutionary triggers for all modern behaviors can be deduced on their worst day and pinpointed on their best. It's a vast field of crops, discovered by the painstaking work of socio-anthropologists and their professional associates.

But we'll need to step it up here. The human evolutionary timeline to which I refer was once a poorly lit alley, even 50 years ago. It is now as well-lit and vibrant as the Las Vegas strip. A struggling human species no longer needs to gamble with its future. We can learn who we truly are for the first time.

What will we learn? We innocent and frightened humans are not flawed. We're gravely injured. We're paralyzed by the immensity of our own existence. We're alone and dying, afraid of the dark, acting every bit of it. The modern, confused human being is not one that we chastise. It's one that begs to be understood, to be talked off the ledge, to be cleared of all crimes and to be saved. Let's go to work.

• • •

Okay. I've named four major human sustainability concerns: climate change, biosecurity, nuclear and disruptive technologies.

If we wait for these models of monetized conflict to view themselves, it makes a fool's errand look like an epiphany. Inattentive blindness sees everything but itself. Therefore, allow me the honor to name a killer. So long as perpetual conflict remains hidden in plain sight, this undefeated destroyer of all past human cultures, and everything we now see in the modern day, has the human species on the ropes, by its inability to view itself, let alone add itself to the list.

Perpetual conflict will end our species. No other rationalization survives.

This message is for you, perpetual conflict. I see you! I know from whence you came, and I will show a petrified world that it's safe to view you, so long as we do not disturb you. You are the unnatural child of The Human Injury, when like and equal beings, born to run, born to help what is helpless, born to assist and to be assisted, gathered into stationary cultures, and were set at one another's throats under models of self-interest. We can understand and quantify the residue of evolutionary paranoia. These ancient and unresolved relics of long-gone realities misinterpret modern threats. The ancient enemies that evolving humans once searched for are long dead, but the mechanisms of brutality alive inside us were never told. Your distorted conclusions are dug into the fabric of modern humanity like fleas. Our helpless species exists now, utterly exhausted and swept along by what we never could see. You'll take out the entire human race to protect the revenue streams you generate.

What you don't know, what you cannot know, is that we understand you. We will not take anything from you, until you feel safe. If we try, we'll be pulverized. You will be allowed to reorganize your deadly inertia. We do not wish to harm you, for that would awaken defensive mechanisms that protect you. Come in off the ledge. Let's talk about this.

• • •

I must assume that the reader is unconvinced. I explain that modern humanity is utterly blameless for the social calamities all around us. You're not buying it. I approve of your unwillingness to settle on existential gibberish. In fact, I counted on it. You want more proof that a helpless modern humanity is completely missing the point. You ask for more evidence, and I deliver, or you won't stay. In response, I give you:

Chapter Two
The Bulletin Of The Atomic Scientists

The people that helm this global vehicle for human sustainability – The Bulletin of the Atomic Scientists - are among the most accomplished people on the planet. If the following is disrespectful, that's a grave misunderstanding. Board members and committee members of this global vehicle for human sustainability hail from business institutions such as Northern Trust and academic institutions such as the University of Chicago. If I were to apply for inclusion on the governing board of "The Bulletin," they would laugh me out of their building. Not because of the credentials I don't have, but because of the common sense I do have.

I don't work for them. I work for eight billion helpless and paralyzed human beings. After I get finished reporting what I can show, trust this, I won't be filling out any applications for consideration on the Board of the "Bulletin." More likely, I'll be shopping for a sweatshirt with a bullseye on it. By writing this, I vaguely endanger myself. This sustainability entity is in no mood for me snooping around.

I have questions for the Bulletin of the Atomic Scientists (herein, the "Bulletin"), questions that will inconvenience this global effort. Thus, the sweatshirt with a bullseye. I bring evidence that The Bulletin can't view itself in any way. I do not bring casual evidence; I bring hard evidence. The Bulletin will be un-amused by the attention I give it. It will respond to my pointed questions not by dignifying them, but by bombarding me with roaring silence.

Why is The Bulletin's silence guaranteed? The ownership that we all require many times over the course of our lives will never arrive. That well-known phenomenon, when inconvenient truth is sentenced to full erasure, gave birth to *The Galileo Effect*. If it could talk, it would know only these words: "Truth is extinguished first and congratulated later." That deadly phenomenon will fall, or humans will fall. The Bulletin is asked the following questions:

1. The Bulletin deploys something called the Doomsday Clock. The clock is set at 90 seconds to midnight. We know that the world has more than 90 seconds, so I ask, what is a cryptic clock doing in the vicinity of a human sustainability effort? What does it mean to an overtly dying humanity when our assumed keepers explain to us that we're "90 seconds" from extinction? Is human life some type of game? The announcement that humanity is 90 seconds to extinction is accompanied by a picture of people chosen to stand proudly on the stage, as human existence is assigned 90 seconds before……. what?
2. If my main premise is correct, that perpetual conflict is unnatural and invariably deadly, and if I easily demonstrate it, might I inquire as to why Wall Street financiers and investment advisors are on the governing board of the Bulletin? A professional financier — a banker — assesses credit risk, not extinction risk. These financial titans lend money to a global model that depends upon perpetual conflict to grow, and they then invest the trillions of dollars reaped by these models of conflict. When a serial killer such as perpetual conflict is missing from the Bulletin's existential threats, that is a damn peculiar thing. My interpretation, moreover, is eminently reasonable. It is not alarmist or accusatory in any way. I demand an answer—in writing.

Below is a sampling of recommendations made by the Bulletin, a vehicle that calls itself the arbiter of human sustainability but acts more like an aggregator of human anxiety. If so, the admission of same would not diminish the group in any way. As you read these recommendations from 2022, let me explain what you won't read. The below recommendations to an overtly dying humanity are a wish list, none of which has ever been done or ever will be done. These empty suggestions are flatly irresponsible—and perhaps deadly. When our doomsday doctors prescribe medicines that can kill us, but do not point to the only killer in the courtroom, what in the Sam Hill is going on here?

As you read these recommendations, your assignment is to locate any suggestion that subsequently occurred. When you don't find any, I'll ask you to note the unusual nature of that very thing. The entire globe is aware of its accelerating freefall to oblivion, while also expecting Fortune 500 CEOs and Wall Street bankers to prescribe solutions for human degradation. If these institutions recognize perpetual conflict as the only known destroyer of human interaction, when will they encourage an end to the warped relationship between lenders and purveyors of conflict? What responsible steward of Planet Earth would not encourage the slow stoppage of revenue streams derived from human conflict?

I'm asking the entire globe to realize the obvious! This group — the Bulletin of the Atomic Scientists — is phoning it in. When bankers and financial advisors counsel a conflict-choked globe on human sustainability, it's vaguely akin to cigarette manufacturers running the American Cancer Society. The irony is considerable. I call upon no less than the most famed comedian in history, Rodney Dangerfield, to add context here. Rodney's hilarious one-liners are a study in bad advice:

1. "I told my psychiatrist that I swallowed a bottle of sleeping pills. He told me to have a few drinks and get some rest." *Rodney's psychiatrist will kill him.*
2. "I told my son about the birds and the bees. He told me about my wife and the butcher." *Rodney should avoid his son when he's feeling blue.*
3. "My fan club broke up. The guy died." *With fans like that, who needs detractors?*

Play time is over. Has the reader deduced where I'm going here? Are we to believe that the Bulletin is unaware that it has banking and investment hawks advising a world choked with conflict on how to survive itself? Are we aware that this international group, the Bulletin is publishing human sustainability recommendations that are flatly impossible? Have we deduced yet that the metaphorical equivalent of The Night Stalker is at the helm of a global effort to rein in serial killers? Why would an active alcoholic counsel anyone on sobriety?

But let's be fair here. The Bulletin never stated that it intends to stop human degradation. It chose the next best thing — to report upon human degradation. If there is no intention to help humanity to survive, then I'm inclined to back off. In return, I ask that this group cease and desist from blowing smoke at the globe in the form of fantasy recommendations, posing as actionable possibilities.

I reiterate that we need leadership. The Bulletin can lead, the minute it deduces its uselessness. Any thief knows how to create a diversion on the east side of town, to burglarize every house on the west side. An overtly dying humanity needs leaders, not enablers.

There's a bonus hidden in this list below. I put it there myself. As you read, pick out the one thing that may have occurred, if this group could view itself. The value that Noēsis could bring to this process is immense. In fact, I created this human sustainability entity – Noēsis – to fill the considerable chasm between current sustainability efforts and workable efforts. To acknowledge my concerns would require the Board to concede its myopia. Don't count on it. Here's the 2022 list:

- The Russian and U.S. presidents should identify more ambitious and comprehensive limits on nuclear weapons and delivery systems by the end of 2022. They should both agree to reduce reliance on nuclear weapons by limiting their roles, missions, and platforms, and decrease budgets accordingly.
- The United States and other countries should accelerate their decarbonization, matching policies to commitments. China should set an example by pursuing sustainable development pathways—not fossil fuel-intensive projects—in the Belt and Road Initiative.
- The U.S. and other leaders should work through the World Health Organization and other international institutions to reduce biological risks of all kinds through better monitoring of animal-human interactions, improvements in international disease surveillance and reporting, increased production, and distribution of medical supplies, and expanded hospital capacity.
- The United States should persuade allies and rivals that no first use of nuclear weapons is a step toward security and stability and then declare such a policy in concert with Russia (and China).

- The Board encourages a global partnership under the banner of species self-preservation. The resulting consortium would seek global consensus that humanity is in crisis. It would not seek capitulation on economic, religious, or cultural affairs. The human evolutionary timeline can be accessed to quantify the weight of evolutionary influence upon modern behavior. The Board warns that perpetual conflict is not survivable. A slow and considered uncoupling from models of conflict would require decades, to allow global revenue streams that rely upon conflict to redirect themselves slowly and painlessly.
- Iran and the United States should jointly return to full compliance with the Joint Comprehensive Plan of Action and initiate new, broader talks on Middle East security and missile constraints.
- Private and public investors should redirect funds away from fossil fuel projects to climate-friendly investments.

Shoulda. Woulda. Coulda. The above suggestions are irresponsible for their utter impossibility. The Board has been asked a question; a put-upon humanity would like a response. Why is the metaphorical equivalent of Jack the Ripper (conflict into perpetuity) not on the Most Wanted list in the global post office that calls itself the Bulletin of the Atomic Scientists?

Only one suggestion from the above list can happen. It's the only suggestion that does not contain the word "should." A global partnership with Noēsis is possible right now. This group would add instant and immeasurable value. The list is long. What's longer, unfortunately, is the list of inconvenient points that we make. The "Bulletin" would view us as a threat. I understand the language of silence. This is the value we would add:

1. Noēsis will lead a global effort to unite all humans under the now-forming banner of species' self-preservation? The effort is already underway.
2. Humans can now know, for the first time ever, the immensity of *evolutionary influence* on modern behavior. Noēsis exists for that very reason, and many more.
3. The evolutionary wrecking balls still alive inside all modern humans misinterpret modern threats. Heavily misaligned responses to these

misinterpretations are linked directly to unnatural conflict among humans.

4. If Noēsis applies for our inclusion on the governing board of the Bulletin of the Atomic Scientists, the Board would likely view it as arrogant, and maybe they would have a point. Still, perpetual conflict adorns the tombstone of all fifty lost empires. The Board's human sustainability vehicle is notably missing its engine.

The Bulletin will not concede. To do so would admit intellectual weakness, or worse, willful blindness. Meanwhile, let's be clear. The Bulletin directs our attention to the many bald spots on the human tire but ignores the big shiny nail of mindless and monetized conflict that flattens the tire every time.

I earlier joked that I should buy a sweatshirt with a bull's-eye on it. [To those who understand the "bull's-eye" reference, I wear an extra-large sweatshirt, and I prefer blue.] The moment this book escapes into the wild, the moment that the Bulletin of the Atomic Scientists promptly ignores it, I'll hardly be shocked. If the group comments on my interpretation (they won't), it would dignify my concern. It would also signal a dim ray of hope for anyone plainly astonished by the spell cast by acquisitive frenzy. The model buys a more comfortable seat on the accelerating bullet train of a helpless and petrified human species, and little more.

My single intention is to bring attention to the FACT that The Bulletin is missing the point! But that won't be enough. Board members who challenge my complaint will sniff that I lack credentials. I most certainly do! I lack the credentials that would allow me to watch for one more second as the most remarkable, unique, and rare species in the Milky Way Galaxy slips further toward an event horizon called Commitment, where both light and a duped humanity share an identical chance of escaping. Behold the plight of modern mankind, leaking oil like a beat-up farm tractor, utterly blameless, divided and conquered, helpless and broken, while the cavalry called the Bulletin has its back.

When the above concerns are not acknowledged, we will not wonder why. Concession is weakness in the ancient battles that rage in the baser areas of the human mind. The Bulletin, far from acknowledging a legitimate concern, will erase me from the face of the Earth by their silence. Why concede the point, when we can start a war, lend money to it, invest its trillions in profits, while deflecting the world's attention to secondary concerns?

Let's pretend that the Bulletin concedes evolutionary influence on modern human behavior, and that the misaligned force we see all around us is its only arrival point. Humanity will glimpse sustainability for the first time that very day! Until then, the "Bulletin" lacks utility.

Just a second here. I explained that what cannot view itself cannot be blamed. There is no blame when problem-solving for human sustainability. There is only observation. My observations are reasonable. They are these:

1. The Bulletin of the Atomic Scientists is an aggregator of present-day global realities. This totality of global situational phenomena is then fitted with a list of suggestions that, even if realized, will not confer added sustainability of any kind.
2. The Bulletin makes no claim that evolutionary influence upon modern mankind's ruinous actions is a factor of any kind. These yearly recommendations describe the symptoms of perpetual conflict, not the cause. Remedial scenarios are suggested; these enjoy a probability of being close to ZERO. If even one recommendation were realized, it would kick the can of perpetual human conflict into a different direction, with just as deadly results.
3. Unresolved compulsions of human evolution that burn brightly inside all modern humans guarantee perpetuated conflict. A known serial killer called perpetual conflict is allowed to rage without interruption, as impossible scenarios issued by the "Bulletin" provide the illusion of remedial efforts.
4. The presence of banking and investment officials on the governing board of the Bulletin defies reason! When banking hawks are entrusted to make human sustainability decisions, hawkish scenarios will ensue. The Bulletin of the Atomic Scientists cannot view itself in any way. Let's be clear—the stoppage of human conflict would collapse the financial markets overnight. This planet has a long wait before Genghis Khan, posing as the "Bulletin," values human life. Admission of same will not arrive. Admission would allow the redirection of conflict-based revenue over the course of decades, or even a century. Current economic models will not tolerate rapid redirection. Humans have monetized their executioner. Our species will back out of this nightmare ever so slowly, or it will destroy us.

I offer an invitation to The Board of Atomic Scientists: encourage the partnership with Noēsis as a commitment to human sustainability. Concede that perpetual conflict is not survivable, in writing. Use these words:

"Perpetual conflict will destroy the human species. Humanity must phase out of ceaseless conflict over the slow march of time, perhaps a century, or it will perish by its inaction. The Bulletin will upgrade its recommendations. We accept in full the inarguable premise called, The Human Injury, as described by Noēsis. Not one of the Bulletin's past recommendations has ever worked. With a heavy heart, we declare that our recommendations are useless. Humankind is accelerating into madness. At the precipice of these choices lies the event horizon called Commitment to Extinction. We will forevermore seek partnerships that phase out of conflict responsibly. We finish by evoking the lyrics of a famous rock song. Thank you.

"Old Charlie stole the handle,
And the train, it won't stop going.
No way to....slow....down."
—Jethro Tull, "Locomotive Breath"

Chapter Three
The Point of Injury

Before we pull apart The Human Injury, let's take a brief journey back in time. By describing the random developments that occurred over multiple millennia, we will view the precise point in time that humanity lost control.

Let's invent a time machine. With it, we can now go back and find the very first historical numbskull who advised human beings to never look back. "Don't Look Back" is a catchy title for a rock song by Boston, but it was terrible advice for evolving humans. Our time machine will allow us to round up every last historical dimwit and fire them straightaway. A sentient humanity, now tasked with evolving to sustainability must look back if we wish to survive ourselves. The reality is this. It's emblazoned on the back cover of this book:

The only light to the FUTURE is that of the PAST.

You may think we don't have a time machine. I think that we most certainly do.

It's known as the evolutionary timeline of the human species. The clarity that socio-anthropologists have achieved in viewing the human evolutionary timeline is a conduit to human salvation. Every modern human being is a walking time machine— a biological cloud computer if you will. We all carry the entire experience of evolving humans in our modern DNA.

The old mantra is *don't look back*, but the new mantra invites all humans on Earth to know who we are, to then glimpse sustainability. If the new mantra could talk, it would invite us in.

A sentient creature that can but will not look back is going nowhere fast. Each of us alive this moment is a biological time machine on legs. We carry the weight of 150,000 prior generations. Call it evolutionary baggage. All of us are buried under the rubble of evolutionary compulsions that long ago lost their utility but continue to fight ancient enemies. Are we helpless modern humans good enough to overpower the immensity of invisible and unquantifiable evolutionary impulses still alive inside of us? Can we overpower an invisible jailer inside of us, to whom we remain fully enslaved? Will this jailer release its prisoners when its job is to destroy them?

Think of evolutionary hallucinations as a gigantic beehive. If I've noticed an evolutionary beehive behind the eyes of all modern humans, then I humbly take credit. How can a lost humanity get this far without having deduced the very nose on its face. Is this the awesome power of inattentive blindness? I think not! The phenomenon is more closely aligned with a septic tank that quietly and insidiously fills and then one day bursts. Modern humanity has awakened to its own madness, but only recently. We can now see the tombstones of all fifty past empires, their corpses riddled with the bullets of The Human Injury. They all perished under the weight of the models we now see today. In this cookie-cutter movie, no one gets out alive.

Most of the evolutionary influence upon modern humanity occurs in areas of the human brain that are consciously unavailable to all of us. We – all humans on Earth—are fighting the same battle against ourselves. It's a battle we can't win. Thousands of knee-jerk compulsions, almost all of them defensive in nature, still search for threats that long ago stopped existing for humans. The conscious human brain in its present moment is a metaphorical molehill that we would ask to control Mt. Everest, in the form of invisible evolutionary drama still alive inside all humans. Such a being is one that we should cry tears of empathy for. This is a bell that we cannot un-ring, should we choose to preserve ourselves. The helpless modern human being arrives today, convinced that it's a scourge on the planet. For what crime? That's precisely the point—there is no crime scene here. This modern generation of humanity fails to do the impossible, to silence the clamor of five million years of DNA-based evolutionary drama. This crime doesn't rise to jaywalking let alone to a death sentence. This is a being that needs to forgive itself.

Fifty thousand prior generations of human beings are packed into every single modern human being that now exists. The immense evolutionary influence upon modern human behavior exceeds our capacity to bear it. The phenomenon is illustrated below. It's the everyday reality of eight billion human beings. Not only are modern humans fighting invisible and crushing evolutionary influence, we're doing so while also acting in a clinically unnatural fashion, courtesy of The Human Injury.

NOĒSIS

EVOLUTIONARY INFLUENCE ON MODERN BEHAVOR

100,000

GENERATIONS

VS.

MODERN

MANKIND

CONCLUSION:

MODERN HUMANS ARE POWERLESS AGAINST INVISIBLE EVOLUTIONARY COMPULSIONS

PRESCRIPTION:

UNDERSTAND EVOLUTIONARY INFLUENCE OR PERISH

SCAN THE QR CODE FOR A DETAILED EXPLANATION OF THIS ILLUSTRATION.

Who do we think we are? What are we asking of ourselves?

Consider the strongest man in the entire world. If we locate this man, we'll look upon him with a sense of awe. But what if we dropped 10,000 pounds of weight on top of him? What was the strongest man in the world just moments ago is now crushed to death under a weight he was ill-equipped to bear. How does the modern human, crushed underneath five million years of invisible evolutionary influence, while also acting unnaturally at the command of The Human Injury, not invite our empathy, rather than our disgust? How did disgust ever get into the picture? I've never been disgusted at a broken arm, have you?

I've proposed a question that could never have been posed prior to 1859, when *On the Origin of Species* hit the shelves. Let's break up this party, and let's all meet in one place, the place where all of these mysteries began.

• • •

That would be the African Congo Basin. Our shared place of origin proves our oneness. No human on Earth was ever born to harm anyone, particularly our own. Because we destroy our own, we will not survive that phenomenon alone, never mind its infinite complications. A sentient intelligence destroys its own and plans for the future as if there is one. An argument that this is remotely survivable – one we don't have – is needed here. Cue the Jeopardy jingle. We'll be waiting a long time.

Human history began in the Central African Congo Basin. Problem-solving for sustainability requires immovable presumptions to be in place. Here is ours: humans evolved on Earth. That's just the way of it. If you don't agree, put this book down. Your right to believe any interpretation for human existence will never be challenged. Thank you for joining me this long. Long life, and good health!

For those who accept evolution, when we say *human* we usually mean modern humans, but the word refers to anyone belonging to the genus *Homo*, which is Latin for "man." Everything that came before *Homo sapiens* can be labelled, "*protohumans*" for simplicity.

Among the earliest groups of protohumans, Handy Man, or *Homo habilis*, ranged across Eastern or Southern Africa between 2.4 million to 1.4 million years ago. *Homo rudolphensis* lived nearly two million years ago near where Kenya is today, while fossils belonging to *Homo erectus* ("Upright Man") have been discovered across all southern Africa, modern-day China, and Indonesia.

Homo erectus was motivated to roam. They had the will! But did they have the grey matter? Unfortunately, no. They tried hard, and the considerable fossil footprint that I just described says as much.

All the while, the mind of natural selection was watching. It noticed that *Homo erectus* was leaking oil in its bid to survive its shortcomings. Natural selection intervened. The additions that she conferred added octane to the fuel it would take for evolving humankind to survive the rigors of the journey ahead.

What did the mind of natural selection notice?

These early humans lacked the sharp teeth and claws of apex predators. Simply put, an apex predator is the alpha species in whatever domain we're exploring.

For instance, the saber-toothed tiger enjoys considerable historical lore, but the big cat was not a welcomed sight for early humans. In fact, anthropologists are certain that no human attempt to affectionately pet a nearby saber-toothed tiger was ever successful because the saber-toothed tiger was an apex predator. In the modern day, a pod of Killer Whales (Orcas) routinely hunt in packs, and feast on a Great White Shark. The Orca is therefore the apex predator of the world's oceans.

These earlier humans lacked other animals' speed, as well, and that just wouldn't do. They made up for these deficiencies with an ever-growing brain that could outthink an apex predator. The natural resources around them were fashioned into artifacts of defense. It was during this time — the early Stone Age of at least two million years ago — that crude toolmaking developed. Those tools include hammerstones, stone cores, and sharp stone flakes. Hand axes and larger cutting tools go back some 1.76 million years ago.

Homo erectus flourished from about two million years ago to approximately 120,000 years ago.

Around 300,000 to 400,000 years ago, another group of protohumans — *Homo heidelbergensis* — migrated from the African continent. The ones who headed north into Europe evolved into *Neanderthals* while those who headed east into Asia became a more recently discovered group called the *Denisovans*.

Many modern humans routinely have Neanderthal and Denisovan DNA. [They have similar temperaments, as well, but that's another book entirely. Evolutionary levity is not my strength.]

Comedy aside, has the reader noticed the lowest common denominator for these early humans? It's central to this book—so central, in fact, that The Human Injury occurred when the following constant was defeated. Namely, our ancestors were always on the move. Human beings are born to run. Springsteen figured it out 50 years ago. Perhaps we should step it up.

The Human Injury is no buzz phrase. It's as real as a migraine, and it's as undefeated as Rocky Marciano. Marciano was a boxer who retired with a 49-0 record. Funny enough, that's approximately the amount of stationary human cultures that once existed on Earth that are now deader than Julius Caesar. It went down like this:

Five million years of human migratory and cooperative DNA arrived at all areas of the globe, around 10-15,000 years ago. It was a biological masterpiece of collaborative, affiliative, cooperative, integrated, sustainable, and loving DNA, an innocent being that would soon be asked to successfully settle down into larger populations. That being would fail every attempt to do just that, precisely and only because it lacked the evolutionary blueprint to negotiate a shock to its well-oiled migratory and affiliative DNA. Simply put, we failed to become stationary beings. We're still failing—and failing so miserably that our mortality as a species is suddenly just around the bend.

This stationary human being would completely break down into something unrecognizable because it lacked the DNA blueprint to act in a manner foreign to its migratory longings. The indifferent nature of natural selection relies heavily upon luck. Evolving humans, in the most literal of terms, were unlucky. This highly advanced being — *Homo sapiens* — ran afoul of a random evolutionary crossover event, where models of acquisitive self-interest first intruded on its linear experience. To be clear, the modern human is not flawed in any way by its choices. It didn't have choices; it only had random events that would overwhelm and confuse it and short-circuit its evolutionary chemistry. That being is us! We failed both then and now by the hangover of unnatural behavior, with its creation of greater and lesser beings within an unnatural hierarchy, foisted upon a being that did not choose it, did not understand it, and that now looks down a shotgun barrel to oblivion by this misunderstood event, despising its own reflection. It's charged itself with everything; it's guilty of nothing. Our collective cluelessness is apparently an existential felony. We don't deserve survival. By that pretzel logic, neither does a lost child deserve to find its parents.

The above empathic interpretation will come up constantly. Many will say that the refrain is repeated too often; I counter that it cannot be repeated enough. The described crossover event from collaboration to acquisition in fact occurred; its hangover is observable and instantly reproducible — the hallmark of any socio-mathematical constant. Every single breakdown experienced by human beings since then, and every single breakdown that will happen tomorrow, is directly and inarguably related to this crossover event. A self-loathing humanity that never deduced its origin, nor ever could it until 1859 is guilty only of possessing no idea how it got here and no idea of how it's leaving. We now watch what we never once understood fall to pieces right in front of us.

We can now value what we've so wrongly accused. Waiting for even another decade to pass without throwing a lifeline to modern humanity, riddled with trillions of bullets fired by The Human Injury, may mean the difference between life and death for an innocent masterpiece called *Homo sapiens*.

Modern humans can now know all of this. The clarity we gain guarantees nothing. The damage already incurred is staggering. We deserve to survive, not to die. Modern humanity will come about, steered by information it never had, but now does. We'll think our way to sustainability with what's now knowable. Stumbling into it—forget about it. Unnatural behavior won't one day magically resolve. We'll own our inertia. We'll abandon our delusions. We'll surrender to our helplessness in full, if only in the name of human survival. We'll redirect, or we will commit to irrevocable oblivion.

• • •

Early humans endured despite radically changing environments. They survived and even thrived when the land they came to know evolved from tropical forests to parched savannah. While human subsets were many, our ancestors shared a uniformly nomadic lifestyle motivated by something in common. The reason that all mammals, including man, roam is for a singular reason: food.

The words *uniformly nomadic* do not have casual meaning. Creatures born to run failed 49 times to be stationary. We're so busy bludgeoning one another, we never considered that basic truth.

THE CURSE OF HUMAN MIGRATION

How can the wonder of human migration morph into the curse of human migration? For the answer, indulge me as we leave Spaceship Earth to explore cosmic phenomena.

• • •

Our Milky Way galaxy beat odds that are thought to be 1 in 100,000,000,000, by offering up an intelligent life form five billion years after the Earth formed. And with this truth, we would call this miracle of creation a curse or a scourge?

When a cosmic phenomenon called *Homo sapiens* appeared on Earth, that event by itself beat all odds. Most humans don't understand the progression that must already be in place before complex molecules of the kind found in a sentient life form can exist. Our appearance on Earth is explained today as having occurred by way of chemical mechanism we do not understand in any meaningful way. Human existence has been assigned one of at least ten mainstream reasons.

As part of an ongoing attempt to coax a faltering humanity into valuing itself enough to preserve itself, I'll refer often to cosmic rarity in this book, and for a good reason. As we develop a better understanding of the universe we live in, we walk back the odds that life exists on any extra-solar planet. Life goes wherever it can, but not in as many places as we once thought.

How did our solar system get here? Astrophysicists have found common ground on how our solar system was formed. Our star, the Sun, was born in a star incubation machine, like all other stars in the universe. The early Sun was surrounded by debris that collected around the sun's immense gravitational field. It had no form, and it was called a solar nebula. Our solar nebula then waited for something to either happen to it (or not happen to it). Something *did* happen, an event that can be called random or divine.

The event may have wiped out our Sun had it occurred closer than it did. The most powerful explosions observed in the universe are known as gamma bursts. These are better understood as time goes by (and they're also observed in distant galaxies). Had our solar nebula been closer to it than it was, it may have been vaporized. As it turned out, we were at a distance where this gamma burst gently pushed our solar nebula into a spinning motion. With too much spin, our shapeless nebula may have flown apart. Had there been too little spin, the solar system may not have formed—but it did. Everything in the universe has a precipitating event. The gentle spinning motion we achieved created the flat disc that forms what is now our solar system.

Why does our solar system continue to spin? Well, objects in motion tend to stay in motion. We would spin into eternity, but our Sun won't play. It's due to run out of fuel in five billion years. Humans will be long gone (and it's a good thing, because you don't want to see this)! Over the course of tens of millions of years, the inner planets will be swallowed whole and vaporized by our dying Sun. It will swell into a "red giant." It will then slowly collapse into a "white dwarf." It will become black and cold, losing all of its energy. Only its dark skeleton will remain.

(More massive stars explode into a supernova, and then collapse into a neutron star. These can spin thousands of times per second. Many of these are now observed in our Milky Way).

For a handy example of what our expanding Sun will look like, look up at the Orion constellation. When you get over its breathtaking beauty, observe the star on Orion's left shoulder (from the observer's perspective). It's called Betelgeuse. It looks reddish and angrier than a hornet. It's nearing the end of its life. Relative to our sun, Betelgeuse's current size would extend all the way to Jupiter. This dying star is 400 light years away. If it exploded 200 years ago, an Earth-bound observer would know it 200 years from now. The star is too far away to hurt us, but its brilliance will one day bathe the Earth in uninterrupted daylight. We will appear to have two sons for several months.

No one knows with certainty the origin of life on Earth. While the human brain is more understood, the nature of human consciousness eludes us. The accepted theories for the emergence of life will further the concept of human rarity. Below are four immovable assumptions:

1. Life exists in the universe. Look in the mirror.
2. All living things on Earth share the present moment.

3. All life forms on Earth are flying on the same spaceship (which we call Earth).
4. Humans share a single origin, the African Congo Basin.

Here are a few theories that may explain the origins of life on Earth:

1. **Extra-Terrestrial Origin**. This theory is broadly called "Panspermia." Rocks have been discovered on Earth that originated on Mars. Rocks are routinely blasted into outer space by cosmic collisions all over the universe. Locally, and because of that, the notion that humans may have originated on Mars has been explored and cannot be disproven.
2. **Biochemical Evolution**. The universe is a gigantic star incubation machine. The number of stars is infinite, but all the newborns perform the same function. Each is a nuclear fusion reactor, where four hydrogen atoms are fused into one helium atom. Excess energy from fusion is released as heat. When the first stars in an early universe died, they exploded. Increasingly more complex elements only then formed. The planets around these stars form randomly, as did ours. Many stars have no planets. For those that do, life forms may or may not appear on these planets. Dependent upon umpteen fully random realities, a planet's relative distance from its star, and the age of that star, will sometimes give birth to the Goldilocks Zone where rocky planets also support liquid water. When that happens, we're not even close to what needs to be true for life to form. Forget about sentient beings because they require molecular complexity that forms only after billions of years. But let's say that life forms develop. In this case, I'd suggest that outer gas giants be available to protect them because gas giants act like bouncers at a bar. They intercept the barrage of asteroids that would otherwise pepper inner planets, thus disturbing any forming life or preventing life altogether. Indeed, our own solar system has four gas-giant bouncers. Jupiter has 94 known moons, most of which were intercepted. (If I ever own a bar, I'll need a bouncer. Jupiter gets the call every time.)
3. **Spontaneous Origin**. In one word, lightning. (Yawn). Laboratory experiments with lightning routinely produced complex amino acids associated with life.

4. **Simple Beginning**. Despite the word, "simple," don't be fooled. This descends from more recent thinking, a fusion of basic physics and a tad bit of philosophy. Just as our lungs breathe and our hearts beat, these theories fully assume the formation of life anywhere it can as a fundamental arrival point of our universe. These theories do not ask why; they ask *why not*? They do not say maybe; they say *of course*! Life is as basic to the universe as oxygen is to our lungs.
5. **Cold Beginning**. When we're through eating a meal, we may refrigerate the leftovers. This will stabilize the molecules in our leftover meals. Just the same, complex molecules may form and remain stable in a refrigerated or frozen state. Just as we revisit our refrigerated leftovers, a later, warmer environment will bring these stable molecules to life by poorly understood (but eventually knowable) means.
6. **Deep Sea Vents**. It's now known that the earliest forms of "advanced" life on Earth were sea sponges. "Advanced" meant that an engine room existed. We call it a brain. Other scientists think that the lowly but beautiful jellyfish may have come first. Note that both are ocean dwellers. Thermal vents are currently the lead suspect in an ongoing investigation to determine the origin of life on Earth. The implications are endless. Any heavenly body, a planet or moon, even one without a nearby sun, is under extreme pressure by forces of gravity. Pressure creates heat. Thermal vents on oceanic planets or moons, even without a sun, even with hundreds of miles of surface ice, have nothing to prevent life from forming near thermal vents on the ocean floor. Further, nothing prevents these life forms from becoming sentient.
7. **The Zoo Theory**. If "thermal vents" are the lead suspect, then the Zoo Theory is easily the most thinly supported one. It postulates that humans were seeded on planet Earth, and we are now under surveillance by benevolent aliens. These eccentric theories labor under the category of pseudo-science. (This term is also highly misunderstood. Pseudo-science is not saved for what cannot be proven. Instead, it contains the sum of theories that cannot be DIS-proven. Other gems of pseudo-science include ufology and multiverse, the latter of which was Stephen Hawking's final choice, prior to his death).

These above possibilities are taught in most universities, which is the metric required to make the list of "mainstream." These reasons are confined to science and pseudo-science, as opposed to religion, but that should never upset anyone who is religiously inclined. There are many who believe that God is — get ready for it — a scientist!

Intelligence on a human level is easily deduced as being unparalleled in the Milky Way Galaxy, and by extension, any galaxy whatsoever. You may have thought otherwise. The more we know, the less probability we assign to find life. If rarity of this kind (a sentient humanity) does not come to understand and cherish its exquisite rarity, we'll continue to treat ourselves and our Earth as we currently do: like confetti, rather than the Rembrandt we are.

Humans deduced evolution 160 years ago. It means that every human being to ever exist on Earth prior to then worked on every mystery without the first clue. When did cluelessness become a crime unworthy of preserving our existence?

The miracle of advanced life, let alone a being like *Homo sapiens*, is assigned a probability of 1 in 300 billion by scientists who may have had too much time on their hands that day. The appearance of a sentient life form is assigned a 1 in 300 trillion probability. It's OK to be fascinated, but I beg all passing astrophysicists to view these numbers as templates for discussion and nothing more. The idea here is to kick-start the human imagination.

Humans achieved sentience somewhere along our migratory journey — not just the ability to think and to question but to ponder our very existence. With these weighty answers not forthcoming, the mind of natural selection noticed the dilemma, and she scheduled an intervention. Our budding human philosophers were gifted with a spiritual default, 75,000-100,000 years ago. This allowed the developing cosmic miracle called *Homo sapiens* to continue pondering its existence, but not to the point of distraction. These spiritually endowed humans gave their existential ponderings over to a Creator. They soon appeared in cave art. This collaborative and affiliative masterpiece called *Homo sapiens* moved as one like a wheel, all its points every bit as vital as the others. This immeasurably rare cosmic phenomenon – evolving humanity— would cross over to what is now Alaska from Northeast Russia only 30,000 years ago. It would trek east and south, and it would eventually complete its journey to all points of the globe 10,000-15,000 years ago.

It would settle down into larger and larger populations. What once sprinted to survive now stood perfectly still. It forgot its purpose that second. What forgets its purpose will logically rust, and rust we did.

Today, a mere 300 generations removed from 50,000 generations of wandering the globe, this miracle of creation arrives today to its ultimate challenge. Its arrival is yet another random comeuppance. This unlucky being is untrained and ill-equipped for the task, to be responsibly stationary. It was a test that human DNA never studied for. Failure was its only option.

Has it hit the reader yet? ***We don't recognize ourselves anymore at any level of our DNA.*** For anyone with dementia, should they one day not recognize themselves, we will cry for them and for ourselves. What once loved and nurtured us now growls menacingly if approached to be comforted and cries in silence when abandoned. It's broken, frightened, alone and dying — an abandoned infant. I'm describing a tragedy of immeasurable depth. It uses the word *scourge* to punish itself. If we ask it why, it can't really say, because a reason to despise itself never existed.

I know why modern humanity despises itself. It doesn't know who or what it is anymore. We've been shamed into privacy by the very ones who were designed to assist us. We're Amelia Earhart, dying a lonely death on an uncharted island. We had the right plane, but the wrong fuel tank.

Is our demise preventable? Technically, yes. Actually? We shall see.

• • •

Let's get back to the cosmos for now. It's more fun there. We'll return to Earth soon. For deluded modern humans, the cosmos is off limits. Window shopping only!

Even as I write, almost 7,000 exoplanets that orbit distant suns have been discovered, mostly by the space bound Hubble Telescope. The data is limited, but accountants routinely extrapolate. Call me a cosmic accountant.

Each star in the galaxy averages one planet each. Ten percent of these planets are rocky and oceanic, but none of these rocky planets are thought to contain life. The other ninety percent are gas giants—cosmic politicians if you will.

You see, the universe we inhabit does not share our romantic requirements that assume a universe teeming with life. Those rocky and oceanic planets have so far proved to be highly toxic. Many are "Super-Earths." They may have double or triple the mass of Earth. If intelligent beings existed, they would be cosmic "shut-ins," unable to leave their planet, due to the energy required to achieve escape velocity. A handful of these rocky planets have 4,000 mph winds on a balmy day. The words that describe human rarity cannot be found. They haven't been invented yet by minds unable to conjure these phenomena, let alone name them.

The number of sentient life forms — on human level and greater — thought to exist in the Milky Way galaxy continues to fall as new data arrives. In fact, humankind may be alone in the galaxy. The universe has no requirements of itself. Life goes wherever it can, but the places it can go are shrinking in number. ET will show itself not to be the dangerous cosmic interloper that our movies describe, but rather, a simple microbe that formed hundreds of millions of years ago, fossilized and forever elusive.

Don't get me started on Mars! For all romantics, you don't want to read this. You may wish to turn away for a second. Mars is unremarkable. Even if Mars is found to harbor fossilized microbes in the ancient lake beds now known to exist or in the liquid water that lurks beneath its north pole, it will prove that a 100,000,000-year window of hospitable conditions that once existed on Mars were a few billion years short of the time it requires to form the complex molecules that are now reading this book.

• • •

The precise path that *Homo sapiens* took in its migratory journey of 275,000 years is now understood with crystal clarity.

Humanity isn't making choices. The ruinous and repetitive actions we take as a species descend almost entirely from residual evolutionary paranoia, in the form of highly misinterpreted modern threats. Modern humans fancy that we're making considered choices. Humans are not making choices any more than a wave chooses to crash. The wave crashes by precisely measurable forces of hydro-physics. Stationary human cultures also crash with precision. Precision denotes a socio-mathematical constant, not a choice.

The current inertia of a crashing modern humanity cannot alter itself without an outside force. Don't look at me! Isaac Newton described inertia.

Post-injury modern humanity is inertia. The situations that our history books describe are variables. Without a process of species' self-discovery, one that we now have, modern humans will continue to delude themselves that we fail by our choices, and the situations that descend from them. We most certainly do not. We fail by the totality of our emotional wounds, all of them descending from The Human Injury.

I can't roll out enough analogies to drive the point home. Try this one on for size. Would the reader please play along as I take us through an exercise in opposite concepts.

The Human Injury is as clear as a bell, but we must first ring that bell to hear it. Once we do, we will never un-ring it. Only the now-abundant evolutionary data for self-discovery will allow humanity to evolve. To gift ourselves even a 1% chance to bust through the technological awakening that's already destroying us, humans will view, we will admit to, and we will surrender in full to our utter helplessness in achieving sustainability.

The injury is open and visible. Don't be offended by the numbing simplicity of the exercise. We've been calling human failure "complex" for centuries. Our clock-like failures are no more complex than the microbe we're currently looking for on Mars. The failure of human cultures over the last 10,000 years is a study in precision. First, consider these opposites:

1. Hot, Cold
2. Wet, Dry
3. Up, Down
4. In, Out
5. Healthy, Sick
6. Sober, Drunk
7. Intelligent, Dimwitted
8. Nomadic, Stationary

Don't look now, but we're hovering over The Human Injury. All readers tolerated the above exercise in opposites so that I can pose a simple question. Let's first agree on this:

IN can never be OUT.
UP can never be DOWN.
DRY can never be WET.

Pretty simple stuff, right? Let's apply it. When we do, behold the exquisite simplicity that I promised.

Only 30,000 years ago, evolving humans arrived to what is now modern-day Alaska after crossing the Bering Strait land bridge, from what is now Russia. These humans already had 50,000 generations of collaborative, cooperative and affiliative behavior sown into their DNA. Collaboration was as natural to evolving humans as breathing. We cooperated, we collaborated, we affiliated, or we DIED.

10.000-15,000 years ago, evolving humans reached every corner of this Earth. Because we did, where was there to go? The human migratory journey was over 10,000-15,000 years ago. What happened next was a series of events that were equally random and logical. These humans, these biological machines of collaboration and affiliation, were commanded for the first time to settle down into larger and larger populations. They were then commanded to abandon the natural hierarchies that maximized the chance of earlier arriving at every point on the globe. Nomadic human DNA was commanded to become stationary.

Remember our above exercise in opposites? Nomadic failed in its attempt to be stationary. Is it that simple? Yes.

These failures would be called tragic and complex by human history books that sold well but knew little. They describe variable "situations." Allow me to encapsulate every history book in the world into one sentence: when the natural hierarchies burned into human DNA crossed over to unnatural hierarchies, the peaceful, dignified, and highly integrated creature called *Homo sapiens* suffered a metaphorical drowning. There it is. Exquisite simplicity.

Since then, this "new normal" was asked repeatedly to inhale deeply in an unnatural environment. It choked on the water immediately, and it never recovered. When stationary human cultures default to models of unchecked financial meritocracy, the clock to social collapse starts that very day. Wherever humans are diminished in any way, and given enough time, these dynamics are uniformly, invariably, and mathematically doomed. [Nothing personal. The math of unnatural behavior always equals zero.]

If I told you that your house was on fire, you'd thank me. If I told you that a dark spot on your cheek needs looking into by a dermatologist, you'd thank me later if it were cancerous. But here, I'm telling an entire species that its current inertia is not survivable under any circumstances, and I'm as lonely as a statue.

If all readers survived the above landslide of the inarguable, let's put a voice to a reality so insidious and suddenly so awful, we're collectively paralyzed. Our instinct is to run from it. All that ends now. Let's turn and face it.

UNNATURAL BEHAVIOR IS DOOMED AT ITS INCEPTION.

I now submit myself for orderly disposal, the fate of all messengers. I can't win this. Let's have a few laughs instead. I mean, why can't oblivion be fun? I'm reminded of a Mel Brooks gag in the movie, *Robin Hood: Men in Tights.* It sounded like this:

> **Sheriff of Rottingham:** "Sire, I have news!"
>
> **Prince John**: "It's not bad news, is it? You know I hate bad news. My day started out so good. I had a good night's sleep. I had a good BM. I don't want any bad news."
>
> **Sheriff:** "Well, to be perfectly frank, it's bad!
>
> **Prince John**: "I knew it! I knew it would be bad. Wait!! I have an idea. Maybe if you tell me the bad news in a good way, it wouldn't sound so bad."

• • •

Unfortunately, we're not in a Mel Brooks movie. It's ALL too real.

The acquisitive modern human being is emotionally exhausted by the frenzy around it. Without a full and unconditional surrender to the suddenly awful realization that human beings cannot evolve to sustainability without a deep understanding of evolutionary influence, we'll do well to cancel all other plans that seek human sustainability. We should shutter every climate change foundation, padlock every last pandemic effort, and call off every planned conference on nuclear non-proliferation. Let's abandon plans to control disruptive technologies, as well.

Instead, let's set one of our AI super-computers to cut the cord of humanity so that we all perish peacefully and instantly.

Why would I prescribe such a thing? I wish it were hard.

We'll fare better in a planned suicide than if we carry on with our current brand of perpetuated violence against our own. Without surrendering to The Human Injury, ceaseless, mindless, perpetual, monetized, and doomed human interaction awaits. That, as they say, is that! Spare me and spare this planet any additional rationalizations that discount the constant equation of post-injury, human desperation.

• • •

If we surrender to our pain, we'll need to monetize the fix, right? No!! We would phase out of conflict over multiple decades. We would turn our attention to the immensity of evolutionary influence upon modern behavior. Humanity in mass retreat from its pain is no mystery. We're overwhelmed, not by the situations we analyze in history books, but by defensive forces inside of us that never got the memo when their ancient enemies ceased to exist.

Let's segue from sweeping generalizations over to specific mechanisms that crowd out any chance to silence the evolutionary beehive behind the eyes of all modern humans. Consider these specific concerns. While you're at it, defeat them.

- Conflict descends specifically from evolutionary hallucinations, forged by 100,000 prior generations of defensive compulsions that hunt with invisible but unabated zeal for threats that no longer exist.
- Evolutionary hallucinations misinterpret everything they see. I stress this. Evolutionary residue exerts influence on everything human beings do, from the most benign of individual undertakings, to the paranoid policies that govern entire countries.
- It gets worse from there. The misinterpretation of Party A, directed toward Party B, summon a misaligned response from Party B. With these evolutionary hallucinations all over us like fire ants, every horse that we send into this cursed race staggers out of the gate, and never recovers. It's a pressure cooker, left alone to resolve itself. It has no

chance. It forms another puddle of perpetuated conflict, and it lays there. We move on to our next adventure, its outcome already assured before it begins.

This broken being is tasked with the impossible, to view what it cannot, and to then defuse the biological chain reaction it sees, by the mere imposition of will. No chance. None.

My above description will be called nihilistic. I might just concur, for a different reason than you think. Nihilism is misunderstood. Nihilism does not imply meaninglessness. Rather, it beckons us to FIND meaning. My nihilistic description is therefore not the epithet you think it is. It just brought to life the first chance that humans have ever had to understand why we do what we do.

Nihilism doesn't condemn us. Ignoring its message is what will destroy us. Nihilism is clarity. Sink or swim—now that's clarity.

The mathematically preordained failures of stationary, acquisitive cultures had no point of reference to explain their existence until the early 1800's, when the best friend we never met, Charlie Darwin, emerged from the birth lottery. These mechanisms of evolution had yet to be deduced. We flew blind for 4,000,000 years. Who do we think we are, and what are we asking of ourselves?

Empathy for a Frightened Child

Stationary civilizations are eventually hypnotized by the bouncing ball called "acquisitive meritocracy." The DNA-based blueprint that would allow stationary sustainability remains missing. An unlucky and innocent child called stationary Homo Sapiens lost thrust, and with it the escape velocity to sustain itself. In this vacuum of knowledge, we struggled over 10,000 years to keep this new technology – stationary acquisition - airborne, and we crashed it every time. There was no other mathematical choice.

The miracle of a successful migratory journey of 300,000 years had no medals waiting for it. The card called "stationary" was dealt to us by an indifferent universe. Stationary and acquisitive humans failed to be the opposite of what their deepest longings beckoned them to be.

Nomadic cannot be stationary until it learns the tune! Modern humanity can now view itself through the lens of great empathy, not of self-loathing. A human species that tries and fails every time to be what it isn't becomes the finest of garments turned fully inside out. An inside-out garment won't play well at a party, and neither does an inside-out modern humanity play well at its current task. A species that no longer recognizes itself is failing at even the simplest of tasks.

We'll change the narrative, not as a casual exercise, but to survive ourselves. Human beings never had faults—only wounds. If this were triage rather than suicide, we would go to the front of the line.

THE POINT OF INJURY

After 250,000 years of collaborative wandering, human beings began to settle. We learned to cultivate crops. Our homes became more permanent, and our populations grew.

For 50,000 generations prior to that, we were a balanced, migratory species that collaborated on all matters. Our community was our family. Life may have been hard, but the facts don't lie. We survived this way for over a million years. Then something extraordinary happened. We changed. Evolution is known to rely heavily upon luck. In evolutionary parlance, therefore, we were unlucky.

We became stationary.

There's that pesky word again — stationary. Recall the analogy I used earlier. An untrained aviator called prehistoric *Homo sapiens* would slam on the evolutionary brakes, and it would soon find itself transported into the cockpit of a Boeing 747. This being would be asked to fly this new technology without a flight manual. This being had no point of reference to define itself. Mr. Darwin's flight manual was still thousands of years in the future. It would be titled *On the Origin of Species*. This being was already in enough trouble, as it slowly descended into behavior that was utterly foreign to its migratory DNA.

Again, metaphorically, this confused being would be asked to land a technology it was ill-equipped to fly. To this day, it wonders why it fails. The evolutionary stage was set for a catastrophe when we settled, and it formed immediately. This being would fail, over and over, with no other outcome mathematically available to it. It would go on to rise from the ashes of guaranteed failures only to be transported back into that same cockpit, asked to perform the same impossible task yet again. Not once, not ever, did this dignified and successful being possess even the basics on how to fly this unfamiliar technology called stationary culture.

What was it about this cultural upheaval — when migratory DNA was asked to do the impossible — that catastrophically derailed evolving humans? The question is also the answer. [Recall first that evolving humans had to this point been uniformly nomadic and that human DNA knew no other way than to be uniformly nomadic.]

With these new demands of stationary cultures in direct conflict with nomadic longings, what could possibly go wrong, other than everything? We devolved. The illusion of civility rolled over us like a rogue wave. If we relaxed our laws today, bodies would pile up way faster than now.

Successful wanderers assumed behaviors of stationary acquisition and fractured clean in half! Natural hierarchy descended into acquisitive meritocracy. Collaboration was discouraged, crowded out, and replaced. A hierarchy of human value intruded. Greater and Lesser human beings appeared on Earth for the very first time. The lesser human being was not merely inconvenienced; it was unable to recognize itself in its very DNA. It rebelled not by choice but by reflex.

Look at this illustration. You're looking at math, not "situations."

• • •

Among the most widely read books on Earth, *On the Origin of Species*, by Charles Darwin is gaining ground. *The Q'uran* and *The Holy Bible* are the top two. That makes perfect sense. Human beings want to BELIEVE. For the first time in human history, we've deduced our origin. We now have the power to BELIEVE and to KNOW. The evolutionary triggers for all modern behavior await our discovery. Their direct role in the mass production of mindless conflicts, both great and small, is easily inferred.

With this Holy Grail in hand, we sentence ourselves to the dungeon of Commitment, not by choice, but by choices we no longer have. Inertia is no choice.

How can a helpless humanity, rudderless until 1859, reach an undiscovered country called sustainability? The rudder is right in front of us—human origin!

ACQUISITIVE MERITOCRACY

Mr. Darwin's flight manua, when it hit the shelves, was of no use to a fully-armed and politically-divided humanity. Humans were and remain openly hallucinating, fighting enemies that never existed, driven by evolutionary paranoia. An innocent humankind was stuck fast in the quicksand of monetized perpetual conflict. The churn of resentment, vengeance, and insurgency that descend from unnatural hierarchy was by then an undetected Stage III Cancer. It entered Stage IV when humankind split an atom. This phenomenon – conflict into perpetuity—collapsed every empire. It stalks and disassembles every human interaction—no exceptions. Show me any diminished human being, and I'll describe its impending collapse. My predictions will come true—not casually, but precisely. A Swiss clock could not do better.

• • •

My analyses inconvenience many, either overtly or invisibly. But be forever warned. If you see yourself as greater, keep it to yourself. No human being on Earth, viewed as a lesser being, will ever abide by it. The lesser being will escape it and has my full encouragement. Dismiss the simplicity of this description at great risk. The **lesser being** will be promoted to **fully equal** in a sustainable future, or that omission alone will sentence humanity to irrevocable commitment. I encourage any alternative argument. I'll wait here, as the Jeopardy jingle plays into eternity.

The innumerable impulses and compulsions inside modern humans can be likened to evolutionary hallucinations. There are no pills for this malady. Instead, evolutionary paranoia was fully monetized by acquisitive stationary cultures. The conflict to which we're fully captive outstrips our ability to abandon it. We're hooked on the heroin of ceaseless conflict. What accompanies addiction is the loss of executive function. Chaos and societal collapse are the lone stations on this long-abandoned railroad. Any human culture that features social and economic models of unchecked acquisitive meritocracy is dead on arrival. Each collapse featured two invariable realities: increasing complexity initiated and exacerbated by inequality. Every dead human culture achieved cartoon-like conflict prior to collapse, with the precision and toxicity of mercury poisoning.

This brings us back to an abandoned and lonely Charles Darwin, whose theories now provide a roadmap to deducing the immensity of evolutionary influence on modern human behavior. These behaviors err on the side of misinterpretation. What misinterprets its environment is hallucinating. Darwin's observations can free us from these ancient prisons, ones that Darwin himself described.

Will we use the information to deduce the immensity of primitive compulsions that activate twenty times faster than thought? If we do not, unchecked evolutionary biology will short-circuit every human response that happens tomorrow, and every day thereafter, until the Earth falls silent. An utterly helpless human species will shrivel and die, blaming everything it sees, while accepting no blame whatsoever. All of Shakespeare's tragedies put together resemble a comic book, when compared to the tragedy of imminent societal implosion. Check that! ONCE imminent. We can escape what we can now see.

THE PRECISE PATH

What specific mechanisms were ignited by the crossover from nomadic and collaborative affiliation to the self-interest of acquisitive meritocracy and eventually acquisitive frenzy?

Did we domesticate wheat, or did it domesticate us? We all agree larger populations yielded more mouths to feed. As migrating humans moved around the world, wheat labored in obscurity—it was a random weed. With the dawn of stationary cultures, wheat catapulted to among the most prolific species on Earth.

Wheat gifted humanity a chance to lessen its struggles. "Economies of Scale" were just up ahead. When they appeared, the human compass cracked for the first time. "Economies of Scale" were the earliest example of societal control. A controlled and efficient humanity was now running against the wind of its own biology. The socio-mathematical constant of human failure was forming. It's all around us, right this very minute—that damn gorilla. It was in utero the day we settled into stationary cultures. Today, it's all over us like a cheap suit. It formed innocently enough, only to collapse into mass production. It looked like this:

1. Hungry humans were compliant humans.
2. Compliant humans were controlled humans.
3. Controlled humans are resentful humans.
4. Resentful humans are vengeful humans.
5. Vengeful humans plot insurgency.
6. Return to #1. Repeat.

It was logical, it was random, and it was ruinous. This babbling brook of growing human resentment grew into a creek, as DNA-based insult piled up. All creeks flow into rivers. The ever-widening river of unnatural hierarchy would soon reach a waterfall, as all rivers must. The violent waterfall of insulted human biology signaled the first collapse of a stationary human culture. There would be many more, each mathematically precise in nature.

At the bottom of those waterfalls, highly agitated and chaotic waters of societal collapse resolved into muted stability. This was the next culture. It noticed the swelling left behind by its failure, but these early settlers were 10,000 years away from deducing their place of origin.

We rose then and we rise now from the ashes of assured failure, not knowing what hit us. The newest river of insulted human DNA would quickly hit those same rapids again, and these would cascade once again into the waterfall. A hapless humanity was now acting unnaturally, relative to its deepest longings. We wrongly blame the GREATER being. But no. We're swept along to failure, not by the wealth of greater beings, but by the resentment of lesser beings.

From here, all forty stationary empires on Earth would act out the same play, unable to foresee the assured tombstone ahead. Our shiny new automobiles, in the form of new cultures that rose from the ashes, were already jalopies the minute they emerged from the cosmic factory. The Human Injury had set in. Its socio-mathematical wreckage is a study in precision. The modern human heart is forever broken, until humanity collectively surrenders to its pain. Leaking existential oil, this helpless being arrives to the present day, calling itself a scourge. We may as well blame our feet for walking.

"The Sea Refuses No River.
We're Polluted Now, But in Our Hearts Still Clean."
- Pete Townshend

What was ahead for The Human Injury could not be imagined. As populations grew, we marveled over our agricultural awakening. Wheat meant that humankind no longer had to roam and forage for food. Planting crops allowed us to build permanent structures, develop better agriculture, and domesticate large animals. We also domesticated our best friends. Dogs are loving animals with a wonderful skill set. Their dispositions were perfect for the job of managing these beasts of burden. Transitioning from a nomadic to a stationary life gave rise to a chain of events, behaviors, technology, societal shifts, and habits that adorn the lavish biographies — and the somber tombstones — of post-nomadic human cultures.

THE FAILURE OF ARGUMENTS TO THE CONTRARY

My apologies go out to any socio-anthropologists that ignore the broader point, who feverishly hunt down minutiae that would challenge my descriptions. I've positioned your work as species-saving, and you would return my gesture by raining down minutiae on the inarguable? I must object. There is no argument whatsoever that can un-ring the bell of The Human Injury. The human crossover (around 10-15,000 years ago) from 50,000 generations of DNA-based migratory/collaborative behavior to a mere 200 generations of stationary/acquisitive models is a fact. Repetitive and machine-like (socio-mathematical) collapse of these models ensued. Self-interested, acquisitive models are in fact unnatural. Concede the broader point and let us transition together to a futurist disposition. It looks like this:

1. Is there a fix within the pages of Mr. Darwin's book? Can the evolutionary motivators for all modern human behavior be deduced? The answer is yes.
2. Can their central and active role in creating and perpetuating conflict among humans be conceded? The answer is yes.
3. Can these models execute a turnabout over the slow march of time, perhaps a century, aided by evolutionary education that recognizes the oneness of all humanity? The answer is yes.
4. Can we emerge as a sustainable species by what we come to know, while allowing these mathematical models to painlessly reposition themselves over decades? We have no choice. We redirect, or we die.
5. Can we wean ourselves off conflict, just as we intend to retire fossil fuels? Yes. Spare me the empty gestures and the unsigned contracts. We do it. Or we die.

Never mind the symptoms of conflict — climate change, biosecurity, nuclear and disruptive technologies. This is a warning to all current sustainability efforts. Any plans that target symptoms, but rationalize perpetual conflict, will fail.

• • •

If we're problem-solving for human sustainability, theory is an orphan. We don't have time to consider anymore theories. Humans have devolved, relative to our deepest longings. Like so many baubles and trinkets, our shiny buildings, our fast cars, our jet planes, and our Armani suits provide the illusion of civility, community, and safety. Who are we kidding?

Are humans intrinsically flawed? Absolutely not. Are we deluded, frightened, paralyzed? Yes. Are we in grave danger by our inattentive blindness? Yes. Are we out of the game? We will blow the dust off Mr. Darwin's book, we will assign ourselves no more right to exist than a slug, or we will disappear without fanfare of any kind.

• • •

Humans were built by nature for natural communion. Consider those who routinely engage in explorative behaviors — such as mountain climbing, camping, or foraging. Moving about in hazardous environments makes collaboration necessary. Unnatural behavior will kill us. No amount of time will ever disconnect the human compulsion to collaborate. That compulsion is wound around the bones of every human on Earth. It's our only hope. Human migration is over, so we will now re-purpose our collaborative longings, not to migrate, but to evolve to sustainability. The current human trajectory is not survivable, by any known or unknown means.

Imagine that we're sitting next to a warm fire at Mt. Everest basecamp. We're discussing the climb to sustainability. We have a Cuban cigar in one hand and a Grand Marnier in the other. We think, *hey, we've got this*! But do we? The economic models sitting next to the fire are unnatural hierarchies. The embers that keep them warm are the like and equal beings that we've diminished for the last 10,000 years and will burn out. Tomorrow, when we begin our ascension to Mt. Everest, our frozen bodies will join all others who attempted a climb they were ill-equipped to start, let alone finish.

> ***"I see dead people. Walking around like regular people.***
> ***They only see what they want to see.***
> ***They don't know they're dead. They're everywhere."***
> \- Cole Sear, *The Sixth Sense*

Hey—don't look at me! Look at the game that I describe. See it, and don't ever unsee it. Teach your children to seek and find the origin of all things. Why waste time? Here's how it went down.

UNNATURAL HIERARCHY; CLASS-BASED MERITOCRACY.

If there can be a "perfect place" to drop yet another sledgehammer of clarity, this might be it. Modern humans maintain the delusion that they live in a "Democracy."

No. We don't. Not even close. The illusion and the reality are separated by light years. That's why you're here. Clarity is just my game. Democracy is just one of many illusions that must end or modern humans will end. All together now! Stationary cultures devolve into financial meritocracy. You always knew it, but perhaps you simply needed someone to plainly shout it out. I'm your man. Let's shout it out to the universe. Are we ready? Once again!!

We Live in a Financial Meritocracy.

I just dispatched yet another torpedo. This one has a name: "Captain Obvious." When delivering these torpedoes of clarity is suddenly a steady job, this returns us to inattentive blindness.

A financial meritocracy is any unnatural hierarchy that diminishes like and equal beings. Show me any unchecked meritocracy. It can be big or small. It might be in the boardroom of a Fortune 500 company. It might be a country club board. It might be in the back kitchen of a greasy spoon restaurant. Its many locations and situations are variables. The constant is precise. Unnatural hierarchies stagger out of the gate and never recover. Given time, with an unchecked inertia, unnatural hierarchy is flatly doomed. All stationary cultures of the past took up acquisition. When *Greater* and *Lesser* beings appeared on Earth, the clock to serial cultural collapse started that second. Please play along as I conjugate the verb *to die*. Watch:

- All stationary cultures on Earth have died under the weight of unchecked meritocracy.
- The current global model of unchecked meritocracy is dying as I write.
- Our species will die at the hand of unchecked meritocracy.

Everyone noticed that I conjugated the verb *to die* three ways, but did you notice that I used the word unchecked three times? If so, you just saw the gorilla! You view the complexity that humans routinely assign to simplicity as the diversion it's intended to be. You've deduced the lone arrival point of unchecked acquisitive meritocracy. You've spotted the tombstone of The Human Injury.

• • •

What phenomenon causes global disinterest in pivotal information? Misinterpreted modern threats are the mother of perpetual conflict. When we unwittingly monetized the serial executioner called conflict, we committed to the demise of forty past empires. I evoke once again the analogy of a sweatshirt with a bull's-eye on it. It's the only fashion choice available to those who fail to consider that their social observations are only as good as the revenue streams they threaten. These very same delusions were suffered by an historical figure of some repute, a man named Galileo Galilei. Galileo was brilliant, but he was situationally naïve.

What does that mean? Galileo thought that his contemporaries would congratulate his truth. And they did congratulate it...but they were not the congratulations Galileo expected. Galileo's heliocentric theory, wherein the Earth is not the center of the heavens, was originally championed by Copernicus. Galileo brought that uncomfortable theory in for a smooth landing, only to be arrested as he emerged from the cockpit of his scientific truth. The compliments directed at Galileo came mostly from prison guards, as Galileo was tortured, starved, imprisoned, and eventually forced to recant his bold conclusions in writing (hence, the Galileo Effect). If it could talk, what would it say? Truth is extinguished first and congratulated later. Should you have any good ideas, ones that might assist mankind in understanding his environment, put the party off. Beware of strangers bearing gifts. Inconvenient truth selects itself for extinction. What did Galileo fail to see? I choose a vaguely comical metaphor to describe a not-so-comical phenomenon:

In 1900, in New York City, word spread of a man named Henry Ford, who was debuting a contraption known as a horseless carriage—a car. That contraption would transform the world within 20 years. Who would ever wish for the automobile to be scrapped? I can think of just one man—let's call him Jim Smith. Jim saw no future in these horseless carriages and silently wished that Henry Ford would contract typhoid fever and pass quietly. Of what crime was Henry Ford guilty? It was the same crime as Galileo: inconvenient truth.

In Galileo's case, the heliocentric theory threatened population control in and around his Pisa, Italy home. The religious population of 1600 Europe was troubled by Galileo's truth. Population control trumps truth any day. When Galileo was tried and convicted, he softly said, "Pero se mueve." But it moves! He meant that the Earth was not the center of the heavens. A gift of Galileo's immensity naively wondered why his truth would ever be suppressed.

This returns us to a devious Jim Smith, who is a metaphor for suppression. The horseless carriage, when it debuted, selected Jim Smith for extinction. Jim owned the largest horse manure cleaning company in New York City. By 1895, New York City's 200,000 horses deposited a whopping three million pounds of manure every day. A world free of horse manure was a world that had endless possibilities available to it. It was also one that had no future for the Jim Smiths of the world.

When human sustainability and horse manure appear in the same sentence, it's not good—and yet it's so real. Acquisitive meritocracy, fully invested in perpetual conflict, is a symbiotic nightmare. We can neither stay engaged with conflict nor can we awaken from it. There's no money in the fix, only in the fissure.

What do these following pairings have in common?

1. Acquisition and Conflict
2. Bonnie and Clyde
3. Nitrogen and Glycerin

To blame these above pairings for their inability to view themselves misses the point completely. None of these pairings saw the concern that they combined to create; these pairings created or continue to create infinite messes. They will not be interrogated, blamed, or judged. Acquisition and Conflict will sense any danger that would pull them apart and would pulverize it into powder. The scoreboard of this unholy marriage, Acquisition betrothed to Perpetual Conflict, birthed by The Human Injury, is:

Acquisition and Conflict: 50
A Helpless Mankind: 0

Every stationary human culture is gone. They did not run their course—no! They were disassembled brick by brick. They each averaged 300 years of existence. Show me any human dynamic whatsoever. If unnatural and unchecked meritocracy is at its core, it's going down.

> "I'll take on anyone, ain't scared of a bloody nose.
> Drink til I drop down, with one eye on my clothes."
> —*Doctor Jimmy*. The Who; Quadrophenia

Remember, there is no blame.

- No past stationary empire deduced evolutionary influence on its behaviors.
- None could ever know that their behaviors were unnatural.
- These cultures were doomed at birth. They suffered immeasurably.
- We see it now. What next?

This is a good time to fire another torpedo of clarity—this one meant to save me, not humanity. Capitalism is the only choice available for sentient creatures.

I say again, acquisitive behavior is our only choice! The modern, broken-down version ((i.e., unchecked meritocracy) is a haunted aircraft carrier that must come about ever so slowly over the course of decades, even a century. It will be asked to surrender nothing until it's ready. It will assume a threat that isn't there. Its acquisitive vice grip and its marriage to conflict will never be challenged nor would any challenge survive. It will deduce its own danger; it will bargain with us, not us with it. It's a bullet train that will either apply its brakes to preserve itself, or self-destruct. Humanity and Unchecked Meritocracy will meet in the middle, or we will die excruciating deaths. When lesser beings become necessary, we have a shot at it.

Sustainability is what transported evolving and collaborating humans to all areas of the globe, over the course of 50,000 separate generations of humans. An undetected, and thereby untreated Human Injury remains the architect of certain human extinction, should humanity fail to view, let alone own the immensity of it. The notion that perpetual conflict among human beings is survivable is preposterous.

• • •

Behaviors that accompanied the stationary reality that followed human migration spawned a series of beliefs, values, and systems that went on to pick off developing human cultures right out of the gate. Human cultures stagger out the gate and don't ever recover. Meritocracy-based human value, in the form of "greater" and "lesser" beings, is a poison pill, a pill that would go down easier if these ruinous, clocklike and now-viewable realities delivered anything in the way of intrinsic happiness among modern humans.

But they don't! That's been proven. We're sick, we're medicated, we're enraged, we're alone, we're frightened, we're broken. This is a tragedy so infinite and awful, so powerful and so compelling, so unnecessary and chilling that it will finally invite and receive the attention for which it desperately begs.

Perhaps you were bored only moments ago when I evoked Galileo. The ownership of these descriptions will never arrive. Anyone surprised or confused by the stark nature of this book should study the chemical mechanisms of addiction. The phenomenon primarily robs its victims of the ability to prioritize. An addict who has two choices, to either submit himself for rehabilitation or press a red button to blow up the planet, will choose the fix and the global coffin that comes with it every time. We're hooked on the endless cigarette of conflict. We've been smoking this cigarette non-stop for 10,000 years.

REDUNDANCY

Past extinctions on Earth — about five billion so far — occurred through varying circumstances, but they had one thing in common. Each of the five billion committed to their demise by an event, whether known or unknown, long before the last of their kind fell. Modern humans look straight past the actual concern – commitment – to the finality of extinction. That's why many people dismiss it with a wave of their hand, saying "I won't be here." These very same people proudly post pictures of their children, grandchildren, and great-grandchildren on social media. If I one day say to my grandchild, "I won't be here," I'll be saying that human suffering is confined only to me.

With fossil records now so rich and diverse, evolutionary theory is now closer to accepted science. Again, if evolution doesn't get it done for you, carry on! Can we agree that evolution solves a lot of problems? If humans evolved on this Earth, and if that timeline is now as robust as the Vegas strip, rather than the dark alley it once was, let's go there and stay there. Let's figure it out for a basic reason. Because we can.

What does redundancy have to do with anything? For evolving and collaborating human beings, there was no redundancy. The sun rose every day, but that's where redundancy ended. That day's adventure for evolving humans was as unknown as it was harrowing. Because Mother Nature had so many weapons, and because we had so few, our only choice was to get bigger, to band together against that day's adventure. Each adventure was different than the last. An indifferent and random universe, one that humans were cast into by the miracle of creation, was one that viewed our daily struggle with brutal indifference. To solve mysteries without clues, evolving humans had nothing other than, well, one another. And one another would prove to be sufficient. The fact that our wandering groups made it to every corner of the globe by 10-15,000 years ago is proof enough that one another was darn good company.

Chapter Four
Modern Humans

Show me a history book, and I'll show you a mathematical constant of such exquisite nature, it makes 2 + 2 = 4 look like an amateur.

Should we fail to know what is knowable, our future streams and rivers may be clear. Climate may be under control. Our weapons may be disarmed. Our runaway technologies may be under wraps. If we did not deduce and offset the evolutionary triggers for all modern behavior, we're all finished here on Earth. These ancient, evolutionary triggers are the lone architects of perpetual conflict.

I welcome any argument that challenges and disproves mine with open arms and concede on the spot. In lieu thereof, no charge that these conclusions are alarmist will be considered. There will be no deflection or redirection from what I've demonstrated. Without this evolutionary knowledge harvested, as part of a much broader effort, I'll sign the death certificate for an inattentively blind humanity right now. Perhaps English poet, John Donne best described the fate of what cannot view itself.

"Send not for whom the bell tolls; it tolls for thee."

Even the best humans among us will lose their executive functions to addiction. There are no outliers. If you believe that you're different, human biology begs to differ. Any of us, if heavily addicted, might launch a nuclear missile if promised an unceasing flow of heroin. They are not choices; these are short-circuited executive functions. Money and power addiction are listed in the Physicians' Desk Reference. They are recognized as psychological maladies. Addiction is inertia, not a choice. No visible means to alter this inertia remotely exists. We're distracted by tens of thousands of climate change efforts, pandemic efforts, nuclear disarmament efforts and artificial intelligence warnings. None of these has ever destroyed a human culture, and none will destroy the current global culture. These four efforts are akin to four members of a basketball team. All of them are important. None of them can be diminished in any way. One team member is missing, the most important one. The center for the human sustainability basketball team does not exist. The center of any basketball team is its unifying force. The four members currently playing are noble warriors indeed! Their unifying force, a deep understanding of evolutionary residue on modern behavior, sits on the bench. It will either enter the game, or its, "Game Over."

• • •

It's customary and proper for humans to mourn tragedies that sneak up and swoop in on innocent parties. The two most visible examples of infinite tragedy in the collective human memory are the sinking of the Titanic, on April 14, 1912, and more recently, the twin towers tragedy on September 11, 2001.

These two events featured several thousand casualties, foisted upon its utterly innocent victims.

The Titanic was a floating metaphor for human dysfunction on every possible level. Its precipitating events occurred long before anyone boarded the vessel, and these later combined to swoop in on an innocent humanity. The immeasurable tragedy of 9/11 was also foisted on its victims, their families, and the forever-shattered sensibilities of any human being with a heart beating in their chests. With the deepest of reverence for the immensity of these tragedies, I evoke them now to cement into place the most basic, and yes, the most terrifying of realities, should humans not come to know what is knowable, and to then prevent what is preventable. I declare the following to be an appropriate warning:

- The "Chixhulub Impactor" committed the dinosaurs to extinction. Multiple decades of residual and measurable effects provided the finality of extinction. The point of commitment, the impact itself, is known in retrospect, 66 million years later.

- An iceberg committed the Titanic to its extinction, but it was a violation of the ship's structural thresholds that ultimately summoned the finality of its extinction. Again, retrospective analysis clarified the details.
- In New York City, on September 11, 2001, two successfully commandeered jet aircraft crashed into and committed the twin towers to their extinction. Precisely measurable effects thereafter assured the finality of the event. Components of a larger conspiracy came to be known, again, in retrospect.
- A viewable event in the human evolutionary timeline — when 50,000 prior generations of migratory and collaborative human DNA became stationary — committed all stationary cultures that followed to their assured failure. The catastrophic insult to collaborative human DNA produced the finality of every stationary human culture after that.
- The precipitating event for modern human dysfunction – The Human Injury - has no mysteries whatsoever. These broken-down models of unnatural hierarchy, in the form of acquisitive meritocracy, have only one mathematical destination. Examination of situational variables may make our machines drive and fly better, but situational examination will not impact the machinery of insulted human DNA. These models of failure have come to call yet again. We will not survive the gorilla, unless we SEE it. Viewing it guarantees nothing, other than conferring a considerably higher chance than what we now have, which is roughly zero.

A counterintuitive reality must now intrude on human bliss. That reality is called clarity. Many will call my above descriptions sensationalist. I counter that all of humankind must be slapped out of its inattentive blindness. You didn't know any of this. You didn't want to. Now that you do know it, is this the end, or is this the beginning?

BACK TO THE STARS

Few scientists have captured the public's imagination as completely as astronomer Carl Sagan.

Many were drawn to Sagan's childlike wonder that was infused with a sense of humility that was so sincere; he made us feel safe to ponder the immensity of our universe, to ponder infinity without fear. Sagan was like a parent who comforts their frightened child. As children, we were paralyzed with fear by things that went bump in the night. Perhaps it was that sound under our beds, or that rustling in our closets. Our fear subsided when our parents arrived and explained the rustling we heard. We were terrified only moments earlier, but with our parents near, we fell fast into a blissful sleep.

One of Sagan's quotes is not particularly uplifting, but scientists have a habit they just can't shake: the tendency to dispassionately report their conclusions. The following was Sagan's dispassionate observation:

"Extinction is the Norm. Survival is the exception."

When I read that, I didn't perceive a sledgehammer—quite the opposite. Here was a gentle soul, fully normalizing a concept from which most of us recoil. That was powerful. All the mysteries in the human experience—the endless rustling in the human closet, the stirring under our beds—could now be observed without fear when Dr. Sagan spoke.

The ultimate mystery for humans may be the mystery of death. Every individual goes extinct. How we comport ourselves in life will impact the circumstances of our departure. I'm delighted to wait for whatever Mother Nature has in store for a self-respecting humankind. What I will not do, however, is sit by and watch the most remarkable, rare, and helpless species in the galaxy select itself for what need not be.

I once met a remarkable woman who spent thirty years working in hospice care. *Surely*, I thought, *hers was among the most difficult of careers*. I was all too happy to be wrong. I had the following conversation with her.

> **Her:** Providing hospice care to the dying was the most rewarding experience I ever had or ever will. Every one of my hospice patients found peace at the end.
>
> **Me:** Every single person found peace? *Every single person??*
>
> **Her:** Every one of them. No exceptions.

Who among us would ever lie about such a thing? Her sincerity introduced me to counterintuitive thought. I would place counterintuitive thinking at my side, and I would never let it out of my sight after that day. It allows me to discuss what is intuitively uncomfortable. It allows me to let go of many cliffs only to discover that I'm only inches from the ground. I bring counterintuitive thought here as a necessary and vital component for problem-solving. A sustainable human species would stop normalizing its pain and instead normalize what makes us uncomfortable. Why? Because of a counterintuitive reality. Calmly viewing the discomfort all around them allowed evolving humans to survive.

Evolving humans quantified the uncomfortable, and plotted survival strategies with what they learned. That allowed our group to stay one step ahead of the discomforts we identified. In fact, every living thing quantifies its discomfort, adapts to it, and thrives.

For instance, if the shirt or blouse that we've chosen is heavily wrinkled, who among us would not simply IRON the garment, or choose ANOTHER garment? All humans on Earth quantify their reality, to only then alter it. Only then can we execute on our hopes and dreams.

I'm describing micro human behavior, but what about macro human behavior? How do modern humans respond to the growing calamities all around us? We see modern humanity dissociating from its pain, thereby exacerbating its concerns. Intuitive thought is our choice, but it delivers nothing of lasting value. What is intuitive fails us, and modern humans live in a failed culture.

We've arrived at the notion that all eight billion of us are a scourge. The ever-crankier modern human being blames everything on someone else and accepts no blame whatsoever. We don't deserve to survive. That arrival point cannot be demonstrated. Humans are designed by nature to assist, to cooperate, to help what is helpless. Collaboration kept our groups alive, but we now see modern humans actively running for the exits. It doesn't add up. We're built to quantify, not to run. Humanity seeking the exits is therefore more evidence of unnatural behavior. With rabid escapism in plain sight, the modern human being has clinically devolved.

The Human Injury is therefore undefeated in its preordained failures. Human beings are devolving away from sustainability even as we must now evolve into sustainability. We will fully surrender to these awful truths. The moment we do, we will glimpse our future. No rationalizations to the contrary can ever intrude, not so long as I breathe, and not so long as my children and yours look to us for hope. They cried out in the night, and we were there moments later. They cry out now, and we're paralyzed.

COMMITMENT TO EXTINCTION

Counterintuitive thought is the new sheriff. If it could speak, what would it say? To avoid human extinction is to discuss human extinction. Let's do it now, with our newly-minted flair for the counterintuitive. Only moments ago, we were running for the exits. Now we know why: we're acting unnaturally. We're alone and afraid of the gathering darkness. Let's quantify it and clarify it. Let's submit ourselves to discomfort so that we might preserve the positives that await us and that will outnumber our conquered fears, ten to one.

• • •

Every past extinction had one or several precipitating factors. There are no exceptions. I'm describing the simplest of concepts: cause and effect. All extinct species followed a precipitating factor. Commitment is the mathematical point past which a return is not possible. Deductively, therefore, that point is knowable only in retrospect.

As a modern-day example of commitment, consider an aviation disaster. A flying machine will now and again crash, and thousands have in fact suffered that exact fate. When these machines crash, the loved ones left behind by these awful events will grieve, and an investigation will commence. That investigation will not be satisfied until the precise cause of the disaster is isolated. The subsequent timeline will isolate the point of commitment, where control of the aircraft was lost to the pilots and given over exclusively to the fates.

Look at the below illustration. The extinction of the Woolly Mammoth is always a handy one. The mammoth became extinct 10,000 years ago. The events that committed this most noble of beasts formed 20,000 years before the last mammoth fell. Use the illustration to understand a phenomenon at the very core of this book. For the Mammoth, **Commitment to Extinction** occurred 20,000 years before the last Mammoth fell. And yet, no human sustainability effort in existence today recognizes Commitment, let alone lists it. That should astonish any reader.

NOĒSIS

COMMITMENT TO EXTINCTION

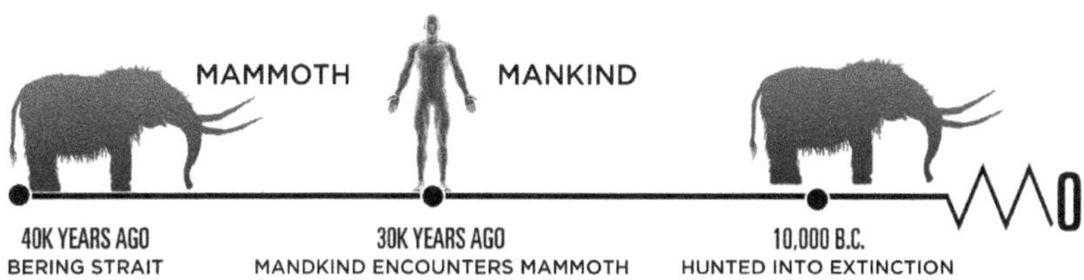

COMMITMENT —20,000 YEARS→ EXTINCTION

CONCLUSION:

- OUR TRUE NEMESIS IS NOT EXTINCTION
- OUR TRUE NEMESIS IS COMMITMENT TO EXTINCTION
- COMMITMENT IS IRREVERSIBLE

PRESCRIPTION:

- AVOID COMMITMENT OR PERISH
- ARBITRARY RESOLUTION IMPOSSIBLE
- ACTIONABLE METHODOLOGY REQUIRED

SCAN THE QR CODE FOR A DETAILED EXPLANATION OF THIS ILLUSTRATION.

This demonstrable fact (that 999 of 1000 life forms that ever existed are gone) defeats any fantasy that human beings will somehow cheat an indifferent universe. The mind of natural selection is all-seeing and all-knowing. It's driven purely by common sense. It hunts for efficiency and then rewards it with longevity. It also hunts just as vigorously for inefficiencies. Like a math professor, when natural selection schedules a pop quiz, you either get it right or get it wrong.

Inefficiencies—such as the unnatural behavior that descends from The Human Injury—have no place to hide. Should any inefficiency occur in nature, and should it cross over to unnatural behavior, cover your eyes. You don't want to see this! If you're in the mood to watch a wrecking ball in action, the brutal indifference of natural selection gets it done every time.

Into this cold reality walks modern humanity, the 50th iteration of failed stationary behavior. 50 empires that came before today were egregiously misinterpreted by historians. They called these mathematically guaranteed arrival points "situational failures." Let me dispose of this nonsense straightaway. Stationary cultures, given time, default to acquisitive frenzy, with no outliers. When they do, class-based meritocracy replaces natural hierarchy. Like and equal beings are diminished in value. These "lesser beings" are no longer free. These now-lesser (biologically insulted) beings arrive at resentment, and then vengeance. The insurgent behaviors that come next provide fodder for history books.

I repeat my earlier reference, word for word. Situational analysis is as useless as a graverobber in a crematorium. By this distortion - that situational analysis has value, modern humanity has no chance to survive, unless we alter the invariably doomed inertia of perpetual conflict.

The will to redirect is forming. The data is also there. The will to survive, in lieu of evolutionary self-discovery, won't get it done. The combination of "will and data" can do it. This most unholy of marriages, conflict and acquisition, must be reasoned with, not to threaten it, but to assure it. We will ask nothing to be surrendered. Resistance to human survival must deduce its own mortality. Until it does, it will misinterpret even our best intentions, and pulverize it.

Because of its helplessness, modern humanity has convinced itself that arriving to the edge of mortality is necessary. These defaults to suicidal ideology are cognitive distortions, reinforced by acquisitive frenzy. Your blame goes nowhere. The lunatic at the wheel of this demolition derby has no name. There is no blame, there is no conspiracy. There is only an amorphous thing at the wheel. It doesn't recognize itself anymore. It seeks absurd pleasures and avoids mountains of pain. It looks to the heavens for absolution of a crime it didn't commit.

A helpless modern humanity is out on that existential ledge, preparing to jump. Its only friend is self-interest. That doesn't compute. Self-interest runs counter to its deepest longings--to belong. It's medicated. it's enraged. It's afraid of a gathering darkness. It calls itself a scourge.

"Dear Father in Heaven...
If you're up there, show me the way.
I'm at the end of my rope.
Show me the way."

—George Bailey, *It's a Wonderful Life*

The story of George Bailey from the Christmas movie "It's A Wonderful Life," serves us well here. George is a man who deserves to live and to thrive but loses his way by the imposition of random developments. As an example, consider an innocent child who watches her parents' divorce, and then suffers years of silent and misguided guilt.

George is a metaphor for modern humanity. We're at the point of a crisis that is so infinite and terrible, even the most scientific among us tends to hand it over to a higher power. Einstein and Newton each became deeply spiritual men, the moment they ran out of explanations.

Modern humanity, like George, ran afoul of random (but logical) events that it can now understand. The simplicity and efficiency of collaborative behavior was and remains baked into our DNA. Our new normal, stationary acquisition was more correctly a Rubik's Cube, so foreign were these changes to nomadic DNA. Modern humanity, awash in false bravado, falls on the very swords it was never meant to forge, spitting in the eye of enemies it never had.

George gave his Rubik's cube over to the universe, but a self-loathing modern humanity is still out there on the ledge of existential despair, preparing to jump.

My gift to this metaphorical George Bailey is a verdict of complete innocence. I beseech modern humanity not to jump; I beseech it to not commit.

We are innocent of our past and present actions. They occurred in a vacuum of knowledge. Every long-gone human empire was dead on its feet long before it fell. The reason is as simple as 2 + 2. Every human empire lacked a point of reference to define itself. Our shared point of origin was not yet known. Every human culture that existed prior to 1859 had not yet deduced evolution. These cultures were thoroughbred racehorses, all of which would be shot dead by the starting gun of self-interested (read: unnatural) behavior.

But now we know the components of mathematically assured failure. For the first time, an innocent humanity is equipped with the data to responsibly fight for its life. Every living thing fights for its life. What fights for its life cherishes its life. Life is so precious to humans that we bargain with the concept of death. Our obsession with the afterlife is proof. Only a being that cherishes its existence would wish its existence to be endless. Into this walks a self-loathing modern humanity, paralyzed and suicidal. There is no greater tragedy than the blind inertia of human degradation.

Let's have the courage to view a degrading human race as we do our individual lives. During our individual journeys, we're faced with peaks and valleys. We quantify our concerns every day. We then plot survival strategies. These strategies are then deployed. These are the actions of every individual's "executive functions." Its actions are precisely comparable to a snowplow. We plow our roads first, and only then do we drive our roads. Should our executive functions fail us (i.e., should our snowplow not be deployed), an immediate crisis of survival is ignited.

What we do next will dictate the outcome. We will advance to sustainability by action, or we will degrade toward commitment by our inaction. Fast forward to 2300 AD. Picture a reality where humanity no longer resides on planet Earth. Let's summon the courage to conjure an unsustainable future. The autopsy would cite human inaction as the precipitating factor, the unwillingness to plow the snow of knowable information that will go on to strand every vehicle we use.

Should my above descriptions conjure indignation in any reader, that indignation is the sort that many use to annihilate the message. I beg the reader to let it in, at risk of further propelling a helpless humankind to the precipice of extinction.

• • •

Did I mention? Dystopia sells! As I write, there are several apocalyptic thrillers playing at your local multiplex. I don't judge it. I liked watching *The Maze* and *The Hunger Games* as much as anyone. Hollywood produces and delivers on these imagined dystopic futures, and we line up around the block to see them. If no new movie is available, we seek a dystopian fix elsewhere. Certain video games sap the souls from our children right in front of our eyes. They dazzle us with graphics, but their plots are less compelling. In some of these games, the players walk in the ruins of the dystopic future that we await. They return to their deepest longings, to collaborate, but not in the way we hoped for. They will collaborate to find other human groups, to destroy them. Human sustainability is in the hands of a murderous lunatic with no name. We've normalized and monetized our executioner. Like the villagers in ancient Pompeii, we await its arrival. Like them, we won't know what hit us. In lieu of redirection, hope for humanity is increasingly reduced to a wish that we're asleep when the wave arrives. It's called paralysis.

Are we uncomfortable yet? I'm not filming a dystopic movie.

> ***"The tragedy is, they know what's going to become of them. They can't seem to do anything about it. But there is another side. You see, I love them."***
>
> —James Hong, as Mr. Wu, *The Day the Earth Stood Still*, 2008

These words were a spoken gesture of affection by an alien being toward an unlucky and helpless humankind, one that deduced its origin in 1859, when Charles Darwin showed up.

"They can't seem to do anything about it." Mr. Wu saw a helpless and frightened being, fully accepting of its fate but just as innocent as the proverbial deer. For both species, the approaching headlights are all we see. Stepping aside is no option for the deer. For humans, I stress that dystopia need not come to call.

But we're emotionally exhausted, aren't we? Acting unnaturally for 10,000 years will do that. What will undo it?

As this journey to either oblivion or sustainability starts, here is my first command. Surrender will not be considered. Humans are capable of greatness! Over a century ago, the New York Times speculated that human flight was one million to ten million years away. The Wright Brothers flew 75 days later. In 2025, we're telling ourselves that we must crash. I say we crash never, not with the totality of what we can know.

Let's get straight to work on the problem again.

Should modern humans continue to devolve into escapist behavior, the ensuing freefall has a sudden stop at the bottom. It's called surrender. It's an unofficial form of commitment. Consider that every weightlifter has a limit! What happens beyond that limit? Collapse. The difference here is that a weightlifter has a spotter; a helpless humanity does not. When the last of our kind falls, another decade will pass before the Earth falls silent, just enough time for the echoes of the human dilemma to fade.

The modern human is hereby ordered to abandon any contention that it can silence 100,000 prior generations of ancient compulsions inside it. Only a slow and considered process, unfolding over decades, one that requires no religious, economic, or cultural capitulation of any kind in the short term, will get it done. If you view human salvation as impossible, that is the language of surrender. It will not be considered.

If human beings achieve clinical commitment, there won't be any sirens going off. We can tell ourselves any lie we want, the idea that we'll figure it out. *Homo sapiens* are corporeal. We're just flesh and blood, and we will not cheat this universe. We will henceforth keep clarity as our constant companion, or our children and their children will awaken to unquantifiable despair. This group that I constructed—Noēsis—isn't about extinction. Noēsis exists to redirect humanity away from clinical (i.e., mathematical) commitment.

You expect your doctor to give it to you straight. I'll give it to you straight right now, or what value do I bring?

- Humans have devolved.
- The point of psychic injury for a once-sustainable humanity occurred

when migratory humans gathered into stationary cultures.

- 150 species per day are selected for extinction for the crime of acting naturally. Stationary humans form unnatural hierarchies. These decayed and collapsed in workmanlike and predictable fashion. These models of global perpetual conflict exist today.

The modern human dilemma is no less "Check" in a game of Chess. A King in "Check" must deduce its reality, or it dooms itself. The King has three choices:

- It can capture the offending piece.
- It can move itself away from danger.
- It can interpose another chess piece between itself and its tormentor.

Behold the efficiency and brutality of the above deductive process. Modern humanity is awash in false bravado, unable to deduce its reality. Check is the only station for unnatural behavior. In response, we feverishly locate and snuff out one brushfire for every three that we ignite. We accelerate into this existential curve, divorced from awareness let alone action, light years removed from sustainability. To deny it means that these ruinous behaviors live another day. Inertia on this level cannot be altered without outside force. Moreover, the altering force for inertia must be strong enough to divert it, to spare the crash landing ahead of it. Only an unblinking self-examination, a full understanding of this most awful reality, will save our species. Only a full surrender will allow the ever-so-slow expulsion of human fantasy. Over a century, with this metaphorical "Stop and Switch" in place, human survival will tick upward from today's probability, which is precisely zero. We will not survive perpetual conflict. No argument to the contrary survives.

LIVE BY THE SWORD...

These below things are equally true:

- Humans will throw down the sword, or we will die. We blame ourselves for monetizing what we did not understand. Our species is innocent of its blindness.

- Humans cannot disengage from conflict, nor are we expected to. Models of conflict must resolve at their own pace. No current revenue stream that derives from conflict will ever surrender. The prime directive for a derailed modern humanity is succinct. Acquire or Die!
- The "Mother of Paradox" emerges. What sustains us has always killed us. It's killing us now. It will kill us.

This madness — humans locked into a perpetual fistfight — puts food on our tables and educates our children. That's why perpetual conflict goes unmentioned. What sustains us will not permit its own exposure. The addict will never curse its heroin. It will protect it with fury. Prior to deducing evolution, we did the only logical thing. We financed a serial killer, perpetual conflict, in plain sight. How could we have known? By the time that Darwin's book hit the shelves in 1859, our fully monetized serial killer couldn't read. It rolled right over Darwin. Every stationary, acquisitive human culture selected itself for extinction the day it was born. These guaranteed and precise failures could not be deduced. Precision denotes a mathematical constant. The Human Injury was still awaiting discovery. That's no longer so. The lone suspect is now identified. The Human Injury – a random evolutionary crossover event - gave birth to perpetual conflict. Case closed.

That's our bogey. Now go to work! Defeat this:

- If humanity does not harvest the data, and commits itself to extinction, there is no return. Global denial, ignoring in full this life-saving data, has only commitment to extinction ahead of it.
- A deep understanding of The Human Injury will allow humanity to come about, to turn ever so slowly, allowing these failed models to realign purposefully but painlessly.

Conflict-fed inaction guarantees existential despair for our children and their children. They will know an unquantifiable hell, akin to Cosmic Dunkirk, but no evacuation strategy exists for commitment.

We know that the above analysis is true, but a not-so-funny thing occurs to us. Our bills are due. Acquire or Die! We get right back into this cursed game, like hamsters on a wheel. We will blame everyone for everything, while rabidly accepting none. We fall back into primitive non-concession, searching for enemies that stopped existing eons ago. We want out. It wants us back—a metaphorical David and Goliath, but David has no slingshot this time. The exits we seek to escape this madness are chained shut.

Don't bother looking for who did this. There is no blame. No one did this. There is only an amorphous thing at the wheel: unchecked acquisitive meritocracy, the gift that keeps on taking, courtesy of The Human Injury.

Chapter Five
The Fall of the Ottoman Empire (Short Version)

Human history books sell the variable situations of human failure. These books sell quite well. There is one concern: everything that happens at the hand of unnatural behavior is a study in precision. It's the same play, with different actors. All the actors die in the end. All of them die; none need to.

An "enemy" is anyone who targets us for harm, whether psychologically or physically. If someone targeted you, today or ever, they acted unnaturally. It felt very personal. It never was. They were driven by evolutionary hallucinations, the serial misinterpretation of their environment, forged by 100,000 generations of unresolved compulsions. They acted in a highly misaligned fashion, in response to an imagined threat.

What humans do within acquisitive cultures is not survivable. All of what you see, every bit of it, is a direct insult to the very core of human DNA. Modern humans awaken every day, in the throes of a fully imagined threat to their survival. Their actions will therefore be heavily misaligned. The rejection response they receive will be even more misaligned, and worse, will perpetuate into eternity.

When the prime directive crossed over from "Collaborate or Die" to "Acquire or Die," the spark of humanity died that day. The starting gun killed the first acquisitive horse, and every horse ever since. The Human Injury is undefeated. Situational analysis of cultural failure has no use whatsoever.

Never mind *The Fall of the Ottoman Empire*. This chapter is hereby christened *The Guaranteed Fall of Unchecked Acquisitive Meritocracy*. Plug in any empire you wish. Turn the lights out when you're finished. Thank you.

Chapter Six
A Plea to All Historians

With the shortest chapter in the book behind us, I once again invite the chagrin of all historians here, but I have no choice. Remember — I serve a degrading humankind. The historian's distaste for my descriptions is guaranteed. I beg forgiveness in advance.

The following story is true. I repeat myself, but I need to, until someone hears it.

I recall a history teacher I had in college — Mr. Bennett. He was a nice enough gentleman if a bit humorless. On one occasion, we were studying the fall of (whatever) empire. At some point, we were required to write a term paper. Like most lads in college, I would often put term papers off until the last day.

As if you didn't!

On one occasion, however, I couldn't pull myself together. I was utterly bored with history books by that time. Knowing that the term paper would be late, I decided to try an experiment. It was risky. It would either be a smashing success of comic genius or an unappreciated bomb. Perhaps my experiment would get points for originality. I would soon see.

I carefully fashioned a term paper book cover, and I made it look extra classy. The professor would later assume that he was in for a great ride if the gorgeous exterior of my term paper was any gauge. I then stuffed the cover full of 40 blank pages. The teacher had no idea what awaited him, but his expectations would be sky high.

When the teacher opened the term paper, I could imagine his reaction. There were 40 pages, but only the first page had text. It contained only 14 words. It precisely said this:

THE FALL OF THE OTTOMAN EMPIRE

- By Tony Wall

Same play. Different actors.

The End.

The professor had a sense of humor. He drew one of those smiley faces, and wrote, "Ya got me!! Well done!" I got a "C" when I turned in the late term paper. The professor loved my humor, but he hated my truth. It exposed the uselessness of situational analysis in post-injury human affairs. In a sustainable future, we would fold failed plays, not re-cast them. We'll do better to read *The Cat in the Hat* over and over. At a mere 268 words, it's a more economical read. Recall the machine that the cat uses at the end to clean up the mess he made, just in time for Mom to come home. If only that machine existed.

As surely as you're reading this, The Human Injury is off somewhere close, concocting its next guaranteed mess. At some point (sooner than we think) the lights will go out, but they will never burn again. A mathematical constant never drifts far away from the stage. It will be called upon over and over, as any constant would.

• • •

Consider this example of the unblinking and indifferent eye with which natural selection views its creations. We discussed dinosaurs earlier, so now let's discuss the extinction of one of the most noble of beasts, the Woolly Mammoth. Even as I write, there are efforts afoot that would resurrect the Woolly Mammoth, using its fully preserved DNA. What could go wrong?

Here's how it went down:

The Woolly Mammoth is thought by many to be indigenous to North America. In fact, the Mammoth's many cousins crossed the Bering Strait land bridge between modern Alaska and Northeast Russia. Humankind would also cross that same bridge, a mere 30-35,000 years ago. Early assumptions that the Mammoth was hunted to extinction exclusively by humans have been replaced by new data. Why let the truth get in the way of a good story? Socio-anthropologists are asked to grant me license, for the sake of the broader point.

The Mammoth weighed five tons, and it was not particularly swift. It was able to lumber along at 10 mph, and that was in a full gallop. This noble but unlucky beast was minding its own business one day around 30,000 years ago. It may have glimpsed something moving in its peripheral vision, something it had never seen before. For one thing, this creature stood on two feet. This was advanced machinery! It also held a spear in its hand. This thing could think!

An added inconvenience was that this creature's demeanor was not welcoming. Rather than asking the Mammoth over for dinner, this thing looked like it would rather eat the Mammoth. And it did just that. Man had only recently crossed the Bering Strait into what is now Alaska. Like the Mammoth, it was working on mysteries without any clues. The Mammoth therefore conferred on Mankind its first culinary clue. The conversation went something like this:

"Look, honey. If we can bag one of those things and keep it outside the cave, it'll stay frozen. You can cook way more, and I can hunt way less!"

That conversation of 30,000 years, along with other natural concerns, would result in an extinct species 20,000 years later. The Mammoth could not replenish its numbers fast enough. We mourn the Mammoth because it was a gentle creature. But natural selection neither mourns nor rejoices. The Mammoth committed to its demise 20,000 years before the last of its kind fell.

Commitment and extinction are two vastly different things.

Back to our hapless modern humans now, unnatural behavior leaves puddles of perpetual chaos lying all over the place, and we then move on. Beings designed by nature for communion now hold one another other at arm's length. It never felt right. It never was.

SOCIO-MATHEMATICS

Everything in the universe, without exception, can be expressed in a mathematical equation. While it excuses nothing, it explains everything. One of the main premises of this book is to shift human thought away from blame, all the way over to empathy, over the course of decades. One way to do that is to prove that what we blame is in fact blameless.

Modern humans blame everything on everyone. The truth is that no one is to blame for anything. How could something so pure (a blameless being) morph into something so awful (a scourge)?

Has it occurred to anyone that blame is a threat? By using four words – blame is a threat – I just foretold the doom of modern humanity. Our models command us to rise from bed every day, and to issue threats, in the form of blame, trillions of times a day. We then scratch our heads, wondering why our blame is uniformly rejected. The legitimacy of our blame is irrelevant. Blame is jettisoned on sight.

When unnatural models of unchecked acquisitive meritocracy crowded out and eventually replaced collaboration, the human need for appreciation was burned at the stake, in favor of an action/reaction churn called "blame/blame rejection." The first mathematical formula of human failure was born.

B + BR = PC. (B)lame + (B)lame (R)ejection = (P)erpetual (C)onflict. Around 10,000 years ago, in stationary cultures, the first perpetuated fistfight between naturally fraternal beings was born.

"We didn't start the fire. It was always burning since the world's been turning."

—Billy Joel

Billy Joel got it dead wrong here! The fire he describes is the evil stepchild of The Human Injury. Human beings are meant to hurt no one, except defensively, in battles that long ago stopped existing for modern humans. I must rearrange Mr. Joel's words to reflect the truth:

"We didn't start the fire. The Human Injury started it. The fire channels the resentment of lesser beings. The fire rejects the blame unnaturally assigned to us."

These new words won't sell records. My apologies to Mr. Joel, who may thank me in favor of the broader point.

• • •

For another look at social math, let's examine the inversely proportional relationship between ethics and consequences. Humans fancy that we're ethical beings. Math explains otherwise. Recall the *Varsity Blues* scandal in which affluent parents relied on bribery and deception to secure a placement for their children in highly selective universities. The scandal arrived at "infinite dysfunction." That's the fate of any unchecked inertia that perceives a lack of negative consequences. The TV personalities that took the walk of shame did not succumb to a vacancy of moral virtue. They succumbed to a socio-mathematical constant. The formula is precisely this:

(NC x 0) + t(x) = (I x Infinity)

Specifically, when (**N**)egative (**C**)onsequences are thought to be **0**, and given enough time (**t**)**x** to distort, (**I**)ndecency is no longer able to be quantified.

This formula explains - inconveniently - that human beings are neither moral nor immoral. We are neither ethical nor unethical. We are neither decent nor indecent. More correctly, we're mathematicians.

Forgive me for taking a wrecking ball to human delusions of moral and ethical virtue, but these delusions, if not exposed and retired, will destroy us. Post-injury, modern human behavior can be quantified with unblinking precision. Show me the probability of detection for any human action whatsoever, whether micro or macro in nature, and I will show you unethical, immoral, and indecent behavior that tailors itself precisely to the described probability. The two numbers will always add up to 100.

I invite any reader to defeat the following.

- If detection probability is 63%, over time, indecency will reach and maintain 37%.
- If detection probability is 9%, over time, indecency will reach and maintain 91%.
- If detection probability is 100%, it means that all eyes are on us. We're good boys and girls, right? Wrong. We did the math. Moral virtue is therefore a delusion, one that awaits better odds of non-detection to adjust itself upward from zero. We wish it would hurry!
- If detection probability is perceived to be zero, human indecency eventually becomes cartoon-like, infinite, and unquantifiable. We eventually revel in indecency where no consequences exist.

The above socio-mathematical constant is not survivable. When modern humans own the fact that our ethics and morals are mathematical equations, inversely and precisely related to perceived negative consequence, humans would become ethical by choice, as a matter of species' survival. Unnatural behavior endemic to The Human Injury, like a cirrhotic liver, breaks down in a precisely predictable fashion. The human species is not only injured by the imposition of an unnatural meritocracy. The breakdowns are precisely measurable, across the entire spectrum of human behavior.

Choices? Nope. Mathematics. Yes—rather precise in nature! Extend it to everything and divorce yourself from emotion. You will never unsee the patterns.

• • •

I once asked a mathematician if he agreed with the above assessment. He didn't know what I was talking about. Perhaps he does now. The collective human lack of discipline is not related to choice. They are the arrival points of collective psychic injury, birthed by a now-viewable crossover event. Evolving human beings were not careless! We moved about with great care, with our heads always on a swivel. We didn't take chances unless there was an actual prize. If the chances in front of us exceeded our capacity to take them, those were chances we didn't take.

I'm using the above example to make a point, one you never considered. Look around you and ponder the totality of human excess, everywhere we look. My God, it may kill us! Human excess vexes us on its best day and disgusts us on its worst. Marvel over the phenomenon of excess, but don't blame it. Our ancient ancestors never took unnecessary chances at risk of death. A modern humanity that makes messes and doesn't clean them up begs our understanding, not our disgust. Who among us would blame a puppy for doing its business on the floor?

Humans make and we then walk away from the excesses we create. Of course we do! We have no choice. What we never once did during our evolutionary journey of 280,000 years, is now something we can't help doing in our unnatural state—stationary, unchecked meritocracy.

How are we expected to clean up messes when the DNA-based blueprint for such a chore never existed? The modern human we blame commands our empathy, not our ire. That being begs to be understood, not chastised. If we're blaming evolutionary mechanisms rather than understanding them, let's blame our feet for walking. I just explained the mathematics of "human excess." We call it a scourge. I call it the only choice available to modern humanity, completely called away from its deepest longings, tasked now with snatching stability from the jaws of assured chaos.

• • •

Will today's culture also fail, like all others before it, by the churn of conflict? **It already has!** The mortality of our species is visible to all humans; that sounds to me like failure. Copycat failures pour out of our history books, and we start it all over again. This lack of logical flow suggests an invisible motivator. It's in plain sight now--the mathematical mechanism behind our collective despair. Until we surrender to it, we cannot evolve. I've asked to be defeated. I ask yet again. Modern human behavior violates migratory, collaborative, and affiliative human DNA at almost every level. Has it settled in yet? ***We don't recognize ourselves anymore***. What doesn't recognize itself must now command our empathy, or the human species will commit to the unquantifiable hell - commitment! - that precedes the finality of extinction.

This book is where human delusions of sustainability go to die. Absent a process of human self-discovery, one we now have, we will not figure it out! That's what all dead empires before this one failed to do. They didn't understand themselves in any way then, and we don't understand ourselves in any way now. "We'll figure it out," No, we won't! We don't even know where to start, but that's okay. I built Noēsis to complete the current team of current human sustainability efforts. I earlier claimed it, and I now repeat it. A sentient species that can but will trace its past has no future. What doesn't understand its failures is doomed to act out ancient battles, waged against long-dead enemies. Humankind will own this reality or be swallowed whole by it.

The failure of these models was once a mathematical assurance, suffered by every stationary culture of the last 10,000 years. That's because evolution had yet to be deduced. Mr. Darwin's flight manual now exists. When will we see this wide-open window to human self-discovery? With it, we can abandon self-loathng. We can know in detail the mathematical mechanisms that drive our "rinse and repeat" madness. Until then, we cannot evolve. We're prisoners to invisible forces, while also acting unnaturally.

The contention that we don't recognize ourselves is a bold one. Let's put a finer point on it. What doesn't recognize itself is not merely inconvenienced. It's fully derailed. It will not right itself. Only heavy machinery can do that. That machinery is the evolutionary knowledge that we never had, but now do, in great abundance.

We will not abandon our children to despair, not as long as I'm standing on this side of the grass.

• • •

Let's look at some more of the mathematical equations for perpetual conflict.

$$Pr(y_i=1 \mid x_i) = 1/(1 + \exp(-x_i ß))$$

In an article titled "Weapon of Math Deduction,"[1] Mother Jones provided a statistical formula proving that as troop levels increase, the chance of a successful political conclusion to any conflict decreases. While compelling, it's also elementary. Escalation and brinksmanship are logical, post-injury arrival points. Many say we're dying. ***We've been dying for 10,000 years.*** The day that greater and lesser beings appeared on Earth, a clueless humanity replaced like and equal beings. Human DNA will not play, my friend! We cannot survive a DNA-based insult.

The equation is also based on research from hundreds of smaller conflicts as well as historical events in recent memory. Sociologists looked at every buildup of troop levels and compared them to successful diplomatic solutions to produce a conclusion and the formula to prove it.

More examples follow! The mathematics of social experiments isn't a new field. I'm not introducing anything revolutionary, just stating the obvious. If you wonder why human behavior can be boiled down to numbers, the truth is obvious. Our behavior is predictable; it's that simple. One more equation proves our propensity toward violence and perpetual conflict.

$$P\_HVC\_NP = EXP(\,) / (1+EXP(\,))$$

This is the formula for the probability of high violence in national conflicts.[2] You didn't know there was such an equation? Violence is emotional, passionate, and seemingly the opposite of predictable, yet there it is in all its clinical glory. We are committed to extinction, and we have the numbers to prove it.

How did we miss this well-known killer: perpetual conflict? Here's a hint: what failed systems admit their failure? Humanity will never admit its failure. We infer it by merely looking around. We humans depend upon conflict. What vanquished every culture prior to our own is at the center of everything we do today. Our executioner is our employer.

The gorilla in the video is the friendly sort. The gorilla that exists will rip your head off to preserve itself. That's not even close to my intent, should I wish to continue breathing. If I'm noticed along the way by the gorilla, I promise only to bring awareness at this early juncture, not to take anything away. Stand down.

In the below quotation from a popular rock song, there is only one word that truly matters. See if you can spot it. It's not a quiz. I'll explain. Consider:

> ***"Exactly when the world is not expecting it.***
> ***Exactly when we're sure of ourselves.***
> ***That's exactly when The Old Gods return,***
> ***And sweep our cities back into hell."***
>
> —Blue Oyster Cult, "The Old Gods Return."

There's more going on in the above quotation than we think. The entire quote is in the spirit of this book. Most will read only the lyrics that describe a conflagration. Look closer now. Do you see it? There's that damn Gorilla! We're not just going into hell. We're going BACK into hell. We've been here forty times, and we'll be here again and again, until the lights go out forever. These mathematical, post-injury breakdowns are doomed to repeat, until we deduce our helplessness, or until they silence planet Earth. These two outcomes have no middle ground.

I wrote it before. The math of unnatural behavior always equals zero.

MACRO HUMANISTS

Observation is just my game. I've always defaulted to observation in favor of judgment or blame. The latter two have no utility in problem-solving, and that's what we're doing here. I long ago began to record these observations as they occurred to me. Large groups of human beings act in an extremely precise fashion, whereas the actions of any individual human are not predictable in any way. It occurred to me that precision of any kind is a mathematical consideration. Mechanisms are things that can be reproduced over and over with predictable outcomes. I was observing socio-mathematical constants. I set out to find the event in human history that ignited the constant formula for human failure.

We are macro-humanists. What do I mean? Well, a micro humanist undertaking is a therapist and a patient. A lot of self-help books focus on what is within the control of individuals.

A *macro* humanist, on the other hand, takes an eagle-eye view of human groups, up to and including the entire human race. As I say, the behavior of any one human being is utterly unpredictable. The behavior of human groups is 100% predictable. Quite a disparity! The implications are stunning. Anything whatsoever that occurs with clock-like prediction is no situation! It's a socio-mathematical constant whose origins are eminently knowable if we look! So, let's look:

Did I find the mathematical constants that I sought? I should say so! That's what modern humans can now do for the first time ever. We can look. We can visualize that crossover event of 10-15,000 years ago when we were commanded to abandon the stunning success of our migratory journey, to then become stationary, to then view our clinical brothers and sisters as lesser beings for the first time ever. That would derail a freight train, and it definitely derailed a thriving humanity. The constant for repetitive human failures fell right in—unnatural behavior.

• • •

Perpetual conflict is everywhere you look—100 feet from you, 100 miles, in each country, in every country. Our economic models are based on warfare, not casually, but centrally. We've promoted the serial killers among us to positions of leadership and we sit back and watch them do what they do best. This isn't a question of individual failing. They're not aliens. They're simply unaware of The Human Injury.

Where there is predictability, we're in the company of a mechanism. .

This goes back to mathematical patterns. I wrote earlier about ethics, morals, and decency. These three concepts are from mathematical, right?

Are they? Defeat this observation:

Human decency ***tends to end*** where action is required.

Let's express it in mathematical terms:

(A x 1) = (D x 0)

When Action is 100% required (A x 1), Decency quietly makes its esccape (D x 0).

If the above formula doesn't move you, you need to look way deeper. When you do, a feeling of existential nausea may set it. The "Decency Formula" reveals that humans are lip-service junkies, and little more.

Mother Theresa is an outlier. Past her, the pickings are slim. We will not survive it, until we see it.

If you view this observation as innocuous, you need to look way deeper. When you do, a vague feeling of nausea may set it. This formula reveals that humans are lip-service junkies, and little more. Actionable humans MAY find sustainability. Lip-service junkies will NEVER find it.

We'll always find outliers in both micro and macro human behavior. With outliers recognized, this formula, to our vague horror, quantifies an ocean of human lip service. Models of self-interest disallow us to do what we were meant to do.

Alas! We're left to produce a bumper crop of decency with only words, which devolves from there into a lonely desert of inaction. We want to be decent, but who has time? Blame and judgement go nowhere when the mathematics of unnatural human behavior can explain everything we see!

A sustainable human species will choose and maintain immovable ethics, morals and decency as a matter of its survival. As it stands, we put our ethics in place, but who are we kidding? An inherently ethical being needs no code. We quickly get busy doing math. Ethics are tailored directly to the probability of negative consequences at any given time. What are the odds of being detected if we are unethical? If those odds are zero, I will show you infinite unethical behavior over time. That's another mathematical equation.

We're mathematicians. We abandon integrity if negative consequences are zero. Conversely, we're good girls and boys only when someone is watching. History proves it. A single person may beat these odds, but a society based on unnatural hierarchies has no shot. It's who we were forced to be, but it's not what we wish to be. Therein lies hope.

We need to own these mathematical patterns, and only then pivot to being ethical by necessity, to survive ourselves. When the crossover event herein dubbed The Human Injury occurred, we were asked to assume behaviors of pure acquisition. That's when our deepest longings were discouraged and crowded out. We turned away from collaboration and toward every man for himself.

For 100,000 generations, it was collaborate or die. Ethical behavior was the way to go. It still is. Nothing has changed. The tools for human sustainability remain! What changed was the path, disrupted by the random crossover event of 10,000 years ago. There were no lesser beings in migratory human dynamics.

When another value was added to the equation—stationary culture/unnatural hierarchy—the mathematics of failure replaced the mathematics of success baked into our DNA. Show me a present-day meritocracy. I'll show you an unchecked one: a constantly failed dynamic. There are no exceptions to a constant. Time is the lone variable in socio-mathematics. It looks like this:

- Diminished humans default to resentment
- Resentment defaults to vengeance
- Vengeance plots insurgency
- Containment efforts eventually arrive at brute force.
- Containment hastens what it intends to slow.
- Imminent Collapse. Time (t) is the lone variable.

There are no variables in the equation, other than time. The equation is precise in its breakdown and its eventual arrival. The escapist proclivities of diminished beings have destroyed every single stationary human culture in the last 10,000 years. The true constant is not acquisitive. Rather, it's stationary. Humans have no DNA-based template for responsible stationary behavior.

Now, look around. Is today's failed global culture a new low? Hardly. It's simply the latest new low. Fifty stationary human cultures of the last 10,000 years achieved this same low. Each would go on to achieve an even lower station called non-existence. They never had a mathematical chance. Failed human interactions did not summon The Human Injury. The Human Injury summoned precise failure.

CHAPTER SEVEN
THE FERMI PARADOX

Planet Earth is a stressful place. Let's take a short trip into outer space. When we understand our cosmic rarity, and when we value ourselves by what we come to know, we have a chance to survive ourselves.

Humans claim to value rarity. Let's find out. I explained that this trip will be short. We cannot linger in outer space, because, you see, we die in space. This one is for any romantic who thinks that human salvation is connected in any way to leaving Earth. We bloom where we were planted. Let's get our arms around cosmic scale.

The universe we inhabit has no romantic requirements. It doesn't care what we wish it to be. It doesn't care what we need it to be. Humans evolved on Earth, under atmospheric pressure. We're not going anywhere. Interstellar travel? Forget about it. Window shopping only! Please accept my deepest condolences.

Let's set the record straight on Mars. If you suspect that Mars is an escape plan, stay with it. When the three richest men in the world (Mr. Musk, Mr. Branson, and Mr. Bezos) are competing to build rockets, the coincidence is notable. These men are building rockets that will transport humans to Mars. Past Mars, our choices end. Interstellar travel is not possible. I'll explain why in a later chapter. But first...

• • •

In 1950, Enrico Fermi sat down for lunch one day. He may have had a quizzical look on his face. After all, he was a bit distracted. Pondering what would become one of mankind's most enduring mysteries can distract anyone. Dr. Fermi suddenly blurted out an unanswerable question, directed at no one in particular. His question was an offering to the entire world:

"Where Is Everybody?"

Dr. Fermi was already the known creator of the first nuclear reactor, a man who was commonly called "the architect of the nuclear age." Dr. Fermi kept regular company with men named Albert Einstein and Robert Oppenheimer. He is recalled as a quiet, gracious, and passionate man, utterly unimpressed with himself.

On this day, Dr. Fermi and his companions were talking about whether multiple other sophisticated beings were elsewhere in the Milky Way galaxy — and by extension, in the universe.

When Dr. Fermi delivered the above question, the most famous galactic contradiction was born. It would later be dubbed the Fermi Paradox. It's still called that. The Fermi Paradox remains, so immense and spectacular are its implications.

Dr. Fermi kept his paradox close to home. By 1950, the Milky Way had already been pegged as relatively unremarkable. The Milky Way is the standard "pinwheel" variety. The forces that create the pinwheel effect can be compared to water circling a drain. A sentient and paralyzed humanity is circling its own drain, so why rub it in? The science of it must be explained in the name of cosmic rarity. The immensity of gravitational forces at the center of these pinwheel galaxies causes the lagging effect we see in the outer arms. In fact, our star (the Sun) was born much closer to the center of the Milky Way. It was then propelled outward by centrifugal force. This is a common phenomenon, unofficially called, "galactic surfing." With few exceptions, the only other galactic shape that occurs naturally is, "elliptical," or egg-shaped.

Dr. Fermi knew then what everyone knows now. The laws of physics and motion are constant in the visible universe.

The existence of other galaxies was not inferred until the 18th century, when French astronomer, Charles Messier noticed a smudge in the optical equipment of his day. The smudge would later be called Andromeda. It was only 100 years ago that Edwin Hubble confirmed its existence. But even then, we still didn't GET it. Andromeda was called an "island galaxy." One hundred years later, there are now 170 billion known galaxies, and at least two trillion inferred. New structures appear daily, as their light finally reaches Earth. No evidence exists that the visible universe has a boundary.

Within this ocean of normalcy exists our star. My apologies to all romantics, but our sun is also an eminently typical one — a yellow dwarf, the main sequence variety. If the awe-inspiring mystery of any star can be called boring, for the purposes of Dr. Fermi's paradox, our galaxy and our Sun are both relatively boring.

Dr. Fermi kept the galactic movie running in his mind. Like Dr. Einstein, Fermi stayed with the problem. Certainly, he reasoned, there must be numerous alien societies living in the Milky Way. By extension, they would have likely spread out, assuming their survival of the technological infancy that 1950-era humanity was also experiencing. That checked out, because humans would go on to reach out to the stars the moment they could. The achievement of sentience—to think and to question; to ponder one's existence—comes only through a fusion of proper evolutionary tools (e.g., an opposable thumb, and a burning curiosity). Sentient creatures solve all smaller mysteries, and only then attack the biggest ones with extra vigor.

With this central premise secured, in Fermi's magnificent mind, things just didn't add up! Something vague but persistent emerged finally as common sense. He put it thus: "...with a modest amount of rocket technology and an immodest amount of imperial incentive, (mankind) could rapidly colonize the entire Galaxy. Within several millions of years, (each of 400 billion stars in the galaxy) could be brought under the wing of empire. Tens of millions of years may sound like a long project, but in fact it's quite short compared to the age of the Galaxy, which is roughly a thousand times more than that."

Before the mathematical implications of these numbers are revealed, I pose my own questions to a degrading humanity. If we were to endlessly knock on any door, how much time would pass before we deduce that no one is home? If someone is home, why don't they answer? Why will they not receive us? One of us would eventually make a bold pronouncement. It's very likely that there is no one home at all!

Let's add context here. If we purchase a Powerball ticket, will we cling to the premise that someone must win, while not considering the probability of 400,000,000 to 1? Because we do purchase these tickets, we understand why the fantasy of human salvation, occurring arbitrarily and just in time remains alive in a notably degrading human landscape, even as the mathematical certainty of unnatural behavior forbids it. Have we pinned our children's futures to the arbitrary occurrence of the impossible?

> ***"And you may ask yourself, where does this highway go?***
> ***And you may ask yourself, am I right or am I wrong?***
> ***And you may tell yourself, My God, what have I done?"***
> —*Letting the Days Go By*, Talking Heads

With the above questions of socio-mathematics deployed, let's examine the mathematics of "The Fermi Paradox."

- After thirteen billion years, 700 times the amount of time for one sentient species not named humanity to traverse and colonize the galaxy at sub-light speeds, we've met precisely no one.
- Even if these imagined beings were to discount interstellar travel as being utterly impossible, that does not survive the paradox! Sentient beings would logically choose the deployment of robot spaceships (read: probes) far in advance of spaceships that transport corporeal beings through the vastness of space.
- If the logical flow is there, then why is the only logical arrival point not there? Why are human beings not dodging a mathematical traffic jam of probes from other worlds that must exist?
- The absence of any evidence whatsoever that sentient life existed elsewhere in the Milky Way galaxy prior to fifty million years ago is talking to us. If the reader is not able to listen to the sound of silence, it is my duty as an arbiter of human sustainability to explain, "The Great Silence."
- There's no one home, my friend! Not maybe. Even one sentient intelligence that existed prior to fifty million years ago would also be a sentient creature that we should have met.
- Even if we deploy the "Mother of Guarantees," Eternal Return, the paradox still survives. Eternal return demands that, where there is infinity, so too must there be an infinite amount of all things. What kind of power could possibly protect the universe from the unblinking demands of existential mathematics? Where is everybody??

THE GREAT FILTER

Since the dawn of the Fermi Paradox, the greatest minds have weighed in. Their many explanations have all been reasoned away. But there is one interpretation that cannot be reasoned away. Its deductive genius is so far impenetrable.

The Great Filter is a sweeping concept. We'll use only a fraction of it here. "The Filter" postulates that sentient intelligence destroys itself early in the process of its technological awakening. Call it "technological infancy." Further, "The Filter" may be a constant, due to the immensity of the task, for any sentient intelligence to survive itself. Now, if you will, follow the bouncing ball to its logical end. Any failed attempt for humanity to survive its crossover to sustainability would be no failure at all. What has never succeeded is not a failure. It's just another day at the cosmic office. Can a staggering humanity cheat "the Great Filter?" If I breathe, so too do I believe that yes, we can. We have our work cut out for us. Let's name it. Humans value rarity. Why do we not value our own? Earth is a spaceship, with many passengers. Perhaps a crew would help.

Let's have some fun! What follows are a few mainstream rationalizations that tell a good story, only to die on the vine of a reality unavailable to them. Make no mistake. The cognitive distortions that follow, far from hopeful, allow us to accelerate into a curve. Humans will not survive these delusions. Here are a few. We're borrowing from popular culture here. These two below rationalizations appeared in one of our favorite movies. Watching it unfold, I was enthralled by the concept of an alien visitor, only to be equally depressed by the conclusions of that same visitor. It sounded like this:

"Your technology is not the problem. The problem is you. You lack the will to change."

- *Klaatu*,

"The Day the Earth Stood Still," 2008

It's a good thing I wasn't present when the above exchange was filmed between fictitious Nobel Prize Laureate, Professor Barnhard (played by Sir John Cleese) and that infamous cosmic interloper, Klaatu (Keanu Reeves). After

Klaatu spoke, I would have summoned the Director. I would have insisted that Klaatu's line be changed to reflect reality. After we finished, a better-informed Klaatu would say this:

"We came here knowing that your species deduced evolution in 1859, when, *On the Origin of Species*, was published.. No human culture prior to that had an evolutionary point of reference to define itself. We cannot blame your species for not knowing what was unknowable until 1859. Typically, you monetized the perpetual conflict that destroys you. That development, while ruinous, is the logical one. Why would we destroy your species? We're here to help.

Our one demand is that that you examine your evolutionary timeline. If not, only oblivion awaits the preservation of your behaviors. It always has, but you never knew why. Be certain to allow multiple decades of time to layer in your evolved choices. Allow your failed models to resolve in orderly fashion. Should you choose to preserve these ruinous models that diminish like and equal *beings*, we will be back, and hell is coming with us. Do you understand?"

With the above changes to the script in place, perhaps I'll be offered the part of Klaatu. My interpretation is an informed one, although Keanu Reeves is better looking. In any event, my intervention will continue. Consider now Professor Barnhard's response to Klaatu:

"All species reach a crisis point. It's only on the brink that people find the will to change. It's only at the precipice that we evolve. That's where we are. Don't take it from us. We're close to an answer."

The Director and I will again ponder the correct answer. This would be Klaatu's response to Barnhard's galactic nonsense:

"You're on the brink now, man! With help like you, your species is helpless. When you reach the brink, the free will that you summon will be missing the data that you require. Your failed economic and social models descend from knowable events, not a choice. You fail, and restart. Your illusion of civility is the very thing that brought me here. Your species, like all other sentient species in the galaxy, is driven by innumerable compulsions from your evolutionary past. They switch off the thinking mind, to protect your species from dangers that no longer exist. The mere *will* to survive, in lieu of a process, has failed every culture in your planet's history. Reintegrate with the compulsions that you deny and think your way out! If not, we will return. Do you understand?"

NOĒSIS

THE GREAT FILTER

HUMANS MAY BE ALONE IN THE GALAXY

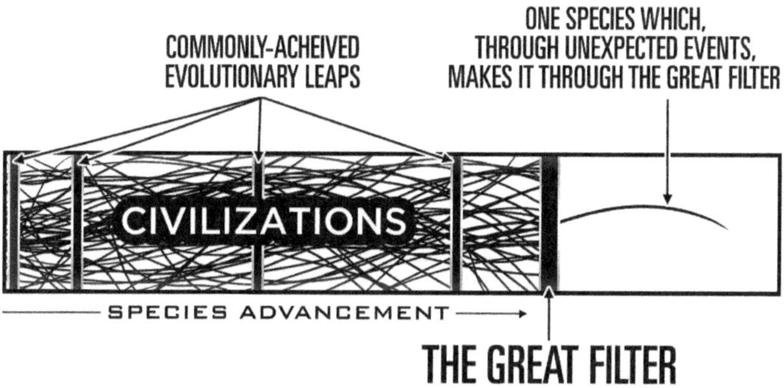

CONCLUSION

EVIDENCE OF EXTRA-SOLAR INTELLIGENCE = Ø

Q: WHERE IS EVERYBODY?

A: THEY'RE NOT THERE

SCAN THE QR CODE FOR A DETAILED EXPLANATION OF THIS ILLUSTRATION.

• • •

I'll never get the part of Klaatu. My speech won't sell popcorn and soda. I understand that. That's why you're reading this book. Humans routinely sacrifice truth in favor of a good story. The freefall created by the sum of these cognitive distortions has a sudden stop at the bottom.

The end of the most remarkable and rare creature is as certain as unnecessary. When human inaction reaches critical mass, it won't be personal in any way. The arrival point is a socio-mathematical one. Commitment! I don't like clichés, but this is a special occasion. Commitment sounds like this. Apologies to James Cameron for the slight alterations:

> *"Water, fourteen feet above the keel. That's five compartments. She can stay afloat with four compartments filled, not five. Not five! She'll go down by the head. There's no stopping it. From this moment, no matter what we do, (Humanity) will founder."*
>
> *"But this ship can't sink!!"*
>
> *"She's made of flesh and blood, Sir. I assure you, she can—and she will. It's a mathematical certainty."*

Titanic had "two thousand, two hundred souls" on board. The Earth may have fifteen billion souls. A soul is a soul. An indifferent universe can't distinguish between souls.

If we fuse together the spiritual and the mathematical, we arrive at a notion that all humans on Earth can live with. Maybe God is a scientist, and a busy one. If only one heavenly body out of twenty billion spawns the mystery of life, that, my friend, is a mathematical miracle. Even an atheist knows what a miracle is. They don't need to believe. They only need to know how to count.

Get your arms around the following information but be kind to yourself if you can't. If we drove a car to the nearest star, Proxima Centauri, we would arrive in fifty million years. If we borrowed a mothballed space shuttle, with a fuel tank the size of Texas, plus plenty of changes of underwear, we'd arrive in

175,000 years. Interstellar travel? Please. We're not going anywhere. We either get it done here, or we're done here.

• • •

Now that we're staying here, let's look further into perpetual conflict. It has many tentacles, all of which are rooted in a misinterpreted threat. Modern humans must abandon the undeserved self-loathing that's currently driving our species into its grave. Species' self-loathing demands that a legitimate reason exists for self-loathing. There are none whatsoever. If humans were making choices, that would be a reason for self-loathing. A post-injury humanity lost its choices the moment it assumed behaviors of self-interest. Your indignation for the behavior of humans who are not you is noted. It's almost certainly misguided. Of the 10,000,000 species currently alive, humans are the only ones acting unnaturally. A self-loathing humanity will not randomly destroy itself. Rather, it will blame, delude, and hallucinate itself into its grave.

I never met an evolutionary hallucination that turned out well. And yet, we suffer them all day, every day. In fact, we're haunted by our evolutionary past. Consider:

- Modern humans misinterpret essentially everything they see. The residue of 100,000 prior generations before ours is a patchwork of invisible compulsions and impulses, mostly defensive in nature, all of them openly hallucinating, searching for, and pulverizing threats that stopped existing tens of thousands of years ago.
- The brutality of these impulses is fully preserved. We can't physically pulverize our enemies anymore, but we can surgically disassemble anyone with hurtful words meant to precisely do that. When you're wiped off the face of the Earth by the words of another, no need to wonder what just hit you. That truck that ran over you was a misinterpreted threat. The details are left for Biologists. The psychology of it is black and white. Something you said was misinterpreted. The human brain does not bargain with threats. You were pulverized by an evolutionary hallucination. Any questions?

- These impulses were designed by nature to locate and destroy anything they don't understand. And these described mechanisms are the vehicles that would transport humanity into a sustainable future? If you think we'll survive invisible forces of abject brutality, I have some land in the Ozarks. I'll sell it to you cheap!

Do you understand what I just did? I foretold the certain death of our species, should we not get ourselves under control, and pronto! If you think Covid-19 was "sudden," you don't know what sudden is. Sudden is when you slip on "black ice." You're strolling along. Twenty minutes later, you wake up in an ambulance. You'll be more careful next time. What if a next time doesn't exist? Humans are toying with an indifferent universe. Go ahead, right this very minute, and snap your fingers. We'll be gone, just like that. Our bravado will reach its apex 10 seconds before the punch that awaits all bullies.

I'm sorry. Have I inconvenienced you? Should I be more clinical? Should I roll out a Princeton study, to win your respect? You haven't figured it out. You're talking to the gorilla right now. Do you see me yet? Your studies and their endless rationalizations allow me to live another day. I own your future.

• • •

When writing about the minute risk of an Independence Day-type alien invasion, Carl Sagan suggested that "…there would be little motivation even for malevolent extraterrestrials to attack the Earth; perhaps, after a preliminary survey, they might decide it is more expedient just to be patient for a little while and wait for us to self-destruct." If I can't be blunt, allow Dr. Sagan to speak for me. I think he said, "We'll be gone, just like that."

Cosmic Rarity (a return)

Random Musings of a Philosopher:

- The modern human will purchase a painting by Claude Monet for $100,000,000. In the ensuing zeal to celebrate, they might step over the bodies of several street-dwellers on the way home. That, my friend, is a tragedy. The bodies we step over belong to the most remarkable and rare species in the galaxy.

- We learn more about our universe in one week than we learned in the entire 20th century. The stunning complexity of molecules required to paint a cosmic masterpiece called *Homo Sapiens* makes a Monet look like a stick figure. Infinite, cartoon-like dysfunction on this level was present in all past empires, right before last call. Now that we know it, let the games end, and let the traffic begin. I do not scold it; I weep for its innocence.
- Our universe is thought to be teeming with sub-sentient life because life goes wherever it can. Sentient intelligence—not so much! Humans reached out into space the moment we could. Are other aliens lazy, or non-existent? The Earth is undetected by alien intelligence and will likely remain so. They don't know we're here, any more than humans can monitor a single microbe on the Yukon.
- If you've deduced that modern humanity is now in contact with The Great Filter, good work by you. Your discomfort is as normal as it is deadly. We have the evolutionary data right now to become possibly the first sentient intelligence to pass through The Great Filter, into sustainability. If we're consumed by The Filter, it will be called a failure by those who turn the lights out. Carl Sagan called it eminently normal. The immensity of the task, to survive itself, has consumed every last intelligence to ever exist. Either that, or something else is going on that defeats mathematics. Except that mathematics is undefeated. Talk about a paradox.

Since life appeared on our speck of a planet, around three billion years ago, upwards of five billion species have already checked into "Hotel Earth," only to subsequently check out. Following a round table discussion on the TV series *Extinction*, the PBS show's website noted that, "Of all the species that have existed on Earth, 99.9 percent are now extinct. Many of them perished in five cataclysmic events.

According to a recent poll, seven out of ten biologists think that we're currently in the throes of a sixth mass extinction. Some say it could wipe out as many as 90 percent of all species living today. Yet other scientists dispute such dire projections. As our panelists debate the issue, they also consider how one species — *Homo sapiens* — may be triggering a modern mass extinction.

THE DRAKE EQUATION

In 1950, as the Search for Extraterrestrial Intelligence (SETI) was forming, American astrophysicist Frank Drake was charged with setting up the parameters. He decided to create a "talking point," one that would enable the conversations that go along with an effort to communicate with ET.

The best way to do that, he reasoned, was to create the "Mother of Conversation Starters," a template that would spark the imaginations of 1950-era humans who were fully engulfed in the UFO phenomenon.

The Drake Equation was born. Dr. Drake would later lament his creation as having been badly misinterpreted. What was meant as a conversation starter morphed into a quantifier of extraterrestrial intelligence. The Drake equation is this:

$$\mathbf{N = R* \times fp \times ne \times fl \times fi \times fc \times L}$$

— N represents the talking point that Drake intended, the number of currently active, communicative civilizations in our galaxy. Drake then got busy eliminating all the "fat."
— R is the rate at which stars form in our galaxy.
— fp is the fraction of those stars with planets.**
— ne is the number of planets that can potentially host life, per star that has planets.
— fl is the fraction of the above that do develop life of any kind.
— fi is the fraction of the above that develop intelligent life.**
— fc is the fraction of the above that develop the capacity for interstellar communication.**
— L is the length of time that such communicative civilizations are active.

**These three assumptions have since been refined, as follows:

1. The Hubble Telescope has now identified 7,000 exoplanets. These are planets that orbit distant stars. According to available data, each star in the galaxy averages one planet. We extrapolate from this that the galaxy contains four hundred billion planets.
2. Ten percent of the planets so far discovered are rocky and oceanic. The other 90% are gas giants. Without gas giants to guard inner planets, those planets are likely to be bombarded with asteroids. Planet Earth would likely not have developed life, had it not been protected by four outer gas giants. Jupiter, for instance, has 94 moons. Most of them were intercepted. Life on Earth has Jupiter to thank. Asteroid impacts would have greatly limited developing life on Earth or snuffed it out altogether.
3. The value (fc) measures how many intelligent life forms would communicate. A non-communicating life form is not a sentient life form. What being that contemplates its existence would not reach out to the stars? We humans reached out to the heavens the moment we could.

Drake's equation determined a probability of between 1000 and 1,000,000 civilizations in the Milky Way that could communicate.

Drake did not wish to be associated with his famous equation because of what astrophysicists and astrobiologists have come to know. Not one of the 700 rocky and oceanic planets so far discovered are likely to have any life, let alone advanced life, let alone intelligent life. When a rocky and oceanic planet has no life on it, all our romantic notions just flew out the window.

In the movie, *Contact*, a young Ellie Arroway wondered if there was life out there. Her father uttered a line that had great cinematic impact, but absolutely no basis in science:

> ***"If there isn't, it seems like an awful waste of space."***
> —David Morse, as Ted Arroway, *Contact*

Of course, you're still not convinced. My relentless nature will fix that. I now give you:

WHY HUMANS WILL NEVER TRAVEL THE STARS

This is yet another point where I feel inclined to apologize. A faltering humanity lessens its already waning chances to survive itself by playing among the stars in its imagination.

KNOCK IT OFF!

Not only is it more evident that ET is probably a fossilized microbe on a distant planet, but even if 10,000 civilizations exist (they don't), we can imagine what they would look like, but we will never, ever, EVER meet them.

I'm a myth-buster. You don't like that part of me, but then again, you don't much like your doctor either. We never see our doctors at family picnics or weddings. What they know about us is a little bit too close for comfort. They seem to get that. As you dislike my message, I can add to the concept of cosmic rarity by explaining why we will never meet the alien intelligence that isn't there. I'm ready to deliver. Are you ready to own it? Let's see:

- Faster than light travel will never be achieved. The very idea that humans will cheat this universe is plainly comical. Accelerating mass to light speed is not possible. Were it to be possible, it would not afford us much of an advantage. Light speed for interstellar travel by humans is plainly useless.
- Don't get me started with "warp drive." It's a human conjuring. The universe, if it had a personality, would be rolling around the floor laughing at the very idea that we will travel faster than its chosen top speed. Let's get a hold of ourselves.
- OK. You're not ready to let it go. Let's pretend that warp drive is possible. In this scenario, we would not be using a method of propulsion. Rather, we would be pulled along in a warp bubble Such a reality, one that cannot be, would create immeasurable violence. Stop with the warp drive!
- Our interstellar vessel would necessarily be one that could never break down. Inconveniently, there are no off-ramps or service stations in interstellar space. Were we to break down, what now?
- Gravity is the constant that holds everything together. It's purely a function of mass. But we'll fix that problem! When we do, we can travel to the stars. We'll create artificial gravity! Umm, no, we won't.

Because we won't, traveling the stars will include floating around all the while. Floating around on long space voyages will dampen our experience considerably. Crossbreeding with aliens on our starship will be harder than ever. No species on Earth can crossbreed, but don't let annoying truths like this get in the way of a good story. Can you imagine my pain?

- Humans die in space. It's a matter of time. Period.
- A manned trip to Mars will occur in 15 years. The required rocketry is the least of our concerns. Keeping astronauts alive for a round trip to Mars requires that almost nothing goes wrong. Never mind the other concerns—isolation, muscular breakdown, radiation, confinement. When we send humans to Mars, and should they suffer a catastrophe and die, the moral and ethical impediments to attempting the same stunt will loom large.
- Elon Musk dreams of placing one million humans on Mars by 2100 AD. Then Mr. Musk bought Twitter. Methinks that Mr. Musk needs forty winks. The man hasn't slept since 2015, and that was a cat nap. One million humans on Mars will be one million miserable human beings. Stop!!
- When we solve millions of physical concerns, and when we then launch our interstellar vehicle, perhaps we'll choose "stasis." Let's kill this idea before it grows. "Stasis" will require the impossible, that the cells in our body obey what we impose upon them. Human cells need to divide at their natural rate. Slowing down the aging process in stasis will kill us. I'm afraid that the human body just won't play! Stasis is the purest fantasy.
- Because of the impossibility of warp drive, our vessels will get nowhere, but at least we'll get nowhere fast! A trip to Proxima Centauri will take 175,000 years with conventional rocketry. A practical matter intrudes on our romantic bliss. Why would we send a vessel on a 175,000-year trip, when a slightly faster method of travel would be only years away?
- Here's what scares NASA worse than anything else: space dust. You read that right. In the vacuum of space, dust does not settle. It was space dust that sickened the Apollo astronauts. You know, the astro-

nauts that never went to the Moon? (Sigh). Whatever floats around that can't be seen will find itself in our lungs. We'll die.

- Our species will survive on Earth or die on Earth. No arguments that make Mars into a steppingstone to interstellar travel will be considered. We can't travel through the vastness of our universe, and let me be clear, our universe doesn't care. When a sentient intelligence like *Homo sapiens* is forced to abandon its home world, tell me, how will things get better from there?
- Should humans relocate en masse to Mars, they'll be dead on their feet. Not because of Mars, but because of The Human Injury .What destroyed all stationary empires on Earth will destroy humans on Mars. It's troubling enough to be unnaturally at one another's throats on Earth. Try the same stunt in a spacesuit.

I advocate for a full retreat from all interstellar fantasy, until such time that we show even a speck of reverence for who we are, for the home we were born on, flourished on, broke down on and now need to recover on. Thank you all!

Chapter Eight
The Creation of Lesser Beings

Imagine you're watching a runaway train from a safe distance. In general, any train is unable to negotiate more than a 2% grade. Any grade whatsoever can run afoul of mathematical guidelines that, if violated, can make for a long day.

The ascent is one thing. But we're not concerned with the ascent. The descent has its own set of unique concerns, mostly related to equipment wear, because precisely measurable forces of gravity use man-made systems to offset them.

Let's say that our descending train has twenty miles to go before it runs out of track or before it collides with another train. Let's further say that our train loses its brakes at precisely the four-mile mark. Let's now imagine the violence associated with a runaway train, with its accelerating inertia fully at the hands of fate.

A runaway train resolves chaotically. While the first responders are cleaning up the mess, the second responders assemble unseen and go into action. In the USA, those responders are called the National Transportation Safety Board, or the NTSB.

The NTSB will look for the precise point along that doomed descent when the brakes failed. In short, they will look for the precise point of commitment, the point from which the train could not recover. When the NTSB finds that precise point, the information will be applied so that the chance of a similar maelstrom is lessened. The investigation will not cease until the circumstances are quantified in whole.

Should any unlucky human be aboard a runaway train, unfamiliar with the train's safety systems, they will spend the rest of their life bargaining with a uniquely awful reality. They will search for any way to stop the train. That effort will later be called as noble as it was in vain.

The post-injury train of humanity, whose stationary disposition is unrecognizable to human DNA, did not lose its brakes. Post-injury humanity never had brakes. Brakes are found only on vehicles that foresee a stop sign. Ten to fifteen thousand years ago, it saw no need to alter a highly evolved, successful, migratory, collaborative being. This being, this "1 in 100 billion" miracle of creation, could not foresee its fate, let alone redirect itself.

Can this miracle of creation forgive itself? Here's the better question: can this miracle of creation deduce its innocence?

Natural selection is a machine of common sense. It paints like Rembrandt, with slow and considered strokes. The crossover event, from collaborative to acquisitive, required natural selection to speed things up. She doesn't work that way. Stationary and acquisitive humankind was now an abandoned mansion, its former splendor able only to be imagined, its descent into ruin assured.

Modern humans are on the latest iteration of a doomed train. The 50th one. But this train is endowed with forethought. It's the first one ever. This train searched for and discovered its timeline. It's the only train in 10,000 years that deduced its origins. It's the only train that can slow its descent. Will inattentive blindness (frenzied acquisition) see the gorilla that it heard about? Or will it keep its nose to the grindstone of disaster? The former is a choice. The latter is inertia. There is no blame.

The above prescription leads us to water. Will we drink? Will we examine ourselves through an unblinking lens? Will we forgive ourselves? Will we declare our helplessness, as we arrive together, as one, to the ultimate task? Will we take the medicine that species' self-examination will gift to us? No other process, no lesser of an undertaking, can or will deliver humanity into sustainability.

Look around you. In any crowd, you'll see people of different shapes, colors, and sizes. Some are native speakers. Others have accents. At first glance, our differences are what stands out. But on a genetic level, the blueprint that makes us human is identical.

There is a less than 0.1 percent variation between any two people on the earth. That's "zero-point one percent." What's not even worth mentioning has somehow been twisted over the last 10,000 years into the very reason to write this book, where I'll ask the smartest species in the room to deduce for the first time what all other species on Earth have no problem deducing.

"Species Recognition." That wasn't so hard, was it?

The innumerable similarities among all humans are never part of our conversations. If we wait for an explanation for how that happened, we're missing the point. Once we became unnatural enemies in or between stationary cultures, we stopped considering our similarities. The Human Injury, unchecked meritocracy borne of stationary behavior, in a horrifying paradox, demands and depends upon unnatural conflict to sustain itself.

No matter the community, the culture or society. In modern macro human behavior, an unnatural hierarchy currently exists. If you want straight math, here it is:

- No unnatural hierarchy has ever survived the resentful beings it creates.
- No unnatural hierarchy is currently surviving the resentful beings it creates.
- No unnatural hierarchy will EVER survive the resentful beings it creates.

That covers the past, the present and the future. What'd I miss? Sociomathematics isn't hocus-pocus. It's as reliable as gravity.

We return to the evolutionary details.

For our now forming stationery and acquisitive cultures, competition for resources slowly discouraged, crowded out and then largely replaced collaboration. Food was a precious commodity and the search for an easier way to secure it was the first brick laid down in the wall of human demise.

The earlier migratory journey, one that began in equatorial Africa 275,000 years ago, was made possible by the deployment of one essential behavior. This would go on to become, over hundreds of thousands of years, the deepest human longing. The immensity of the task for human migration created the ingrained trait that would sustain us. Indeed, what's burned into the wood of human DNA can never be erased. It's impossible to reverse this very necessary factor in humans. This element is seldom seen today. I'm talking about collaboration. One hundred thousand generations winds around the bones of newly born infants of the modern day.

Modern humanity gives endless lip service to collaboration. We reflect upon it. Some groups vaguely embrace it as fundamental. In any event, The Human Injury made this vital characteristic a mere footnote. Collaboration exists now in the third person. It's as if collaboration is in the next room, like a patient waiting for a doctor. Collaboration gets no respect, no respect at all! If that sounds like Rodney Dangerfield, let's let Rodney explain the now-diminished phenomenon called collaboration. This line was one of his favorites.

> ***"My psychiatrist isn't helping me at all. The other day,***
> ***I told him that I felt invisible.***
> ***He said, "I can't see you now."***
>
> —Rodney Dangerfield

Human collaboration did not simply level the playing field. It was a necessary element of survival, every bit as required as breathing. Evolving and migrating human groups either collaborated or died. I stress this for effect. Collaborate! Or Die! This strongest of human longings was—and remains—imprinted in our DNA. It allowed us to survive for hundreds of thousands of years in relative peace, but globe-trotting all the while. Within migrating human groups, we viewed each other as like and equal. We were not lesser; we were uniformly necessary. There was no illusion of ownership. The present moment was all that we owned, and that was darn good enough. There existed not even a budding concept of what the term *acquisition* meant.

Acquisition did not lie dormant inside the developing human species, awaiting its chance to one day emerge as the new normal. There is no acquisitive default sown into human DNA. The frenzied acquisitive behavior we see today, borne of unnatural and unchecked meritocracy, was foisted upon a peaceful, dignified, and unwilling humanity. What was our crime, other than running out of places to migrate, and logically forming stationary cultures? No one knew then what we know now. The crossover from collaborative migration to stationary acquisitive meritocracy collapsed everything to come after that. Without an altering force, the tombstone that's now forged for modern humanity will feature only four words:

UNNATURAL BEHAVIOR CANNOT PERSIST.

Human cultures never once succumbed to their faults, only to their wounds.

• • •

As prehistoric mankind made its way around the world, it had no clue what DNA was. The knowledge that each person in our migrating group was equally necessary came naturally, not because it was hip or trendy, but because it kept our groups alive. Stationary human cultures collapsed into unnatural hierarchies. On that same day, we diminished our own for the first time. We committed to perpetual warfare. The fuse to repetitive, mechanical, and precisely predictable human failure was lit. We remain today, at endless war, not against our enemies, for these never existed. We're at war with our own biology.

The first human culture that established roots also required structures that rise from them — not just physical structures, but social structures that are utterly divorced from our nomadic longings.

THE FIRST DOMINO

The dynamics of stationary cultures (i.e., unnatural behavior) led us into conflict... perpetual human conflict. The conflicts we create are often subtle. They can also be as dangerous as a nuclear standoff. If those two extremes seem wildly unrelated, they most certainly are not. Just as the laws of physics and motion are constant all over the visible universe, the mechanisms of perpetual conflict display constant similarities, no matter their origin. The very same mechanism that starts a playground fistfight—a misinterpreted threat—will one day end the most remarkable and rare species in the Milky Way galaxy.

Perpetual conflict began when a phenomenon called *resentment* first barged into the human experience on Earth. When stationary cultures assigned greater and lesser status to like and equal beings, we broke down that very day, immediately, catastrophically, and irrevocably. Such was the ferocity of a DNA-based insult. Unchecked meritocracy features unnatural hierarchies.

The benefits of natural hierarchies are as strong as unnatural hierarchies are doomed. In fact, hierarchies in nature were central to the formation of complex organisms. It's the very cornerstone of natural selection. If a biological machine has centralized commands (such as the ones we receive from our brain), it runs more efficiently. Natural selection creates and rewards efficiency. Imagine what nature will do to any inefficiencies that it happens upon. Imagine the brutal efficiency with which an unamused Mother Nature will sweep away the sandcastle rationalizations of modern humans.

By contrast, a negative photograph called unnatural hierarchies formed in stationary cultures. It produced complex societies and fed urban expansions. Once an unnatural hierarchy of human value upon the human experience, the trouble swept across an unwitting and innocent humankind like the extinction event tsunami I described in the *Introduction*. The unnatural dynamics of acquisitive behavior brought The Human Injury to life. Collaboration and acquisition repel one another. Nomadic human DNA doesn't recognize the self-interested dynamics of acquisition.

Is it that simple? Yes, it is if we let it be. The complexity we assign to simplicity of this nature allows these models to live another day.

(I've been asked to submit these conclusions to academia. I flatly refuse but not for the reasons you think. Can academia help? No. Academicians in competition will nit-pick a collapsing humanity into its grave. I will not submit this work. I'll do even better than that! I invite an open debate on these issues, in public. I will receive no offers. No academician will risk his or her reputation by denying the mathematics of certain human demise. To do so would be to imply that human inertia might resolve arbitrarily. Don't even TRY it).

THE SCOREBOARD

No matter the structure, from slavery to unions, a "lesser" being has never tolerated anyone who claims to be "greater." The intolerance of lesser beings is a dull ache that feeds every revolution, whether cultural, political, religious, or economic.

If that were not troubling enough, this hyper-acquisitive modern churn will not tolerate being noticed. What it will not view is also what dooms it. Collaborate or die! collided head-on with Acquire or die!

How do we treat The Human Injury? We know the crossover event in human history that brought it to life. We know what makes it tick. But how do we unwind this thing?

A conflict-laden humanity requires (and now has) the evolutionary data and the courage to view itself. It must not fear the near-term loss of anything. Having deduced its arrival to oblivion at the hands of these wrecking-ball behaviors, it can then realign its interests with limited or no pain. I caution that any threat to these ruinous mechanisms will awaken them. No revenue stream of any kind will surrender itself, nor does it need to. If it requires an entire century to return from diminished beings back to necessary beings, let's hope we have that century.

I'm compelled to tell a true story here. It's meant to describe the power of the word "diminished."

When I was around 16 years old, I worked in a restaurant on the Ocean City, NJ, boardwalk. It was your classic burger joint. I was in heaven. I flipped burgers while also girl-watching. Burger flipping made me cool. Watching pretty girls pass by was the cherry on the sundae. I was a normal kid.

The short-order cook in the back kitchen was a misanthrope, a mean-spirited and serially nasty man. He was demeaning on his best day; he could crush your spirit on his worst. Giving certain individuals authority of any kind will not end well, and this man was that individual.

But really, how much damage can a mean-spirited short-order cook do? I would soon find out. I mostly stayed out of his way.

A Mexican employee worked there. His name was Paco. He was a quiet, gracious, and hardworking person. I may not see a work ethic like his ever again. He was silent and efficient. I knew a few words of Spanish, so I engaged with him. I liked him; he liked me. In this young man, I saw a gentle soul who wouldn't hurt a fly.

The short-order cook didn't see that. I saw a human being, but the cook saw a meal. He rode Paco hard. He diminished and demeaned Paco's every effort, for sport. Finally, the axe fell. The cook fired this gentle-hearted soul. It broke my heart. My instinct was to go back into the kitchen before Paco exited the rear door so I could wish him well. I turned the corner, and there he was, just ten feet from the door. He looked at me, as he choked back tears.

I looked back at him. As Paco wept, he urinated into a giant soup kettle on the floor beside him. That kettle contained the soup that the restaurant patrons would order for the rest of the evening. Had I not turned that corner, and had I not seen what I saw, no one would be the wiser. Everyone who ordered that soup may have later fallen sick.

Paco saw me turn the corner, and he recoiled. I nodded my head to him. I said two words: "Yo entiendo." "I understand." Paco nodded to me, he thanked me, then he turned and walked out.

Did I just tell you a throwaway tale that was lost in time? Or did I just explain the illusion of power that a greater being thinks he has versus the very real power that a diminished human being *does* have? You know the answer. Wherever there is chaos and destruction, start your subsequent investigation by locating the most diminished human being. Then locate his "superior." When you do, you just found the silent and long fuse that leads to a big bomb called insurgency.

UNCHECKED INJURY FURTHER DECAYS: THE PATH TO SLAVERY

The transition to slavery for post-injury mankind was relatively rapid. Like a virus, it had the capacity to hide inside the body and render its host violently ill within minutes. The Human Injury did the very same thing to a now-degrading humanity.

At first, this newly minted acquisitive concept called "ownership" was confined to a hut and a shiny new sword. You needed your home and the means to defend it. From there it was a short trip to thousands of acquisitions, and then millions. Advanced machinery—pre-injury migratory humans—were fast becoming a cheap card table that could bear only the weight of the house of cards on it, and nothing more.

The more an individual possessed, the more his possessions reinforced the illusion of greater being*s*, and the more it aggravated the indignation of lesser beings. Slavery was not a human choice. It was more correctly the arrival point of unnatural inertia.

In the modern day, slavery looks like an immoral aberration, a tragedy from another time and place. Unfortunately, slavery hasn't gone anywhere. In fact, no matter how you position your views on our fellow man, the result has always been the same. I earlier described the mathematics of ethics and morals. We enslaved other human beings only after deducing that negative consequences for doing so had fallen to zero. You guessed it: we kept slaves because we could. It would later be called complex. It never was.

No human being judged to be lesser by another will ever abide the lower station assigned to it because sentient creatures must be free and equal! By contrast, if we put a mighty lion into a cage, it will pace angrily back and forth. It wants out, but its bravado won't last. If we return an hour later, the mighty lion is now laying down, grooming itself. It's not going anywhere. Even this most majestic of beasts gets that, and quickly.

But show me a "lesser" human being. It will never tolerate the lesser station assigned to it. Forget about uprisings, insurgency, and wars. These are the variables that descend from a constant. The insult to any human being made to be lesser is not situational. Its chosen actions are variables. The DNA-based insult is a biological constant. An imprisoned human being awaits the chance, not to demur like the lion, but to explode from its prison and to strike back against its jailer.

In ancient civilization, it was common for greater beings to enslave lesser beings. Greater beings treated lesser beings with all the respect and due process afforded an insect. At this point in The Human Injury, where consequences for abuse were zero, we mercilessly punished or murdered what was lesser. By then, post-injury humanity openly reveled in moral and ethical vacancy, whenever lesser beings were helpless, and when consequences were thought to be zero.

We can find an egregious but typical example of the human tendency to accelerate indecency in the absence of consequences in the six Crusades of the middle centuries. These atrocities, which were Vatican-sanctioned events, were recorded by the medieval version of an embedded reporter. By the last Crusade, indiscriminate slaughter was described as ankle-deep in blood. The Vatican had removed the First Commandment, and with it, any possibility of consequence.

• • •

As our societies progressed and the first laws imposed, did we end slavery? Of course not. It simply went underground. If the consequences were thought to be 25% probable; indecency adjusted itself downward to 75%.

In the first half of the 19th century, popular newspapers advertised slave auctions. Today, human chattel is bought and sold along the dark corners of the internet. Chattel refers to legal ownership, but again, who are we kidding? No one who enslaves human beings thinks in legal terms—they only think in terms of negative consequence probability. We won't find enslaved people working on plantations or wearing chains. In the modern world, enslaved people suffer in silence, out of sight. An unsustainable humanity is only as good as the innumerable secrets it keeps. Decency need not toil in secrecy.

The modern slave model has many variables, all driven by the same constant. Where consequences are none, where fear rules and where enough time passes, the most well-intentioned effort devolves into infinite debauchery. There are thousands of diminished and devalued humans ten miles in any direction from where you're sitting right now.

The key lies in finding the courage to dig deep into how we value each other. If you do that, you'll also face the sobering concept that we judge indiscriminately, but unknowingly! Wherever unchecked meritocracy is observed, what is "lesser" undermines and destroys any system. The breakdown, moreover, is a study in precision, as follows. Set your watch by it:

- Resentment gives way to vengeance.
- Vengeance plots insurgency.
- The failing model deploys compensatory behaviors.
- This compensation arrives at brute force.
- Brute force hastens what it intends to preserve.
- Collapse is imminent; time is the variable.
- Greater beings abscond with whatever they can; lesser will never forget.
- We restart the model.

Left to repair itself (i.e., unchecked), our present global models are dead on their feet, not casually removed, but light years removed from sustainable.

On the very day that the first human being was diminished in value relative to other humans, due to unnatural meritocracies that descend from class-based models of acquisitive self-interest, a metaphorical torpedo was dispatched to the bow of collaborative human DNA. The torpedo scored a direct hit. The bow of a peaceful, integrated, fully sustainable and highly dignified creature blew clear out of the water. This modern vessel, despite its gleaming exterior, is a ghostship at its core, in its bowels, where the lesser people huddle. Call this compartment "steerage." Call these lesser beings the ragged people, the dishwashers. Call them whatever you will, so long as you also call them the assured demise of every human interaction that ever occurred or ever will.

Those we diminish will answer to the names we call them, but they will never become what we make them, so long as they breathe, and so long as the greater being deludes itself otherwise. They do not rebel by choice but by biological reflex.

As responsible stewards of a great gift, our call to action in the modern day is simple. Change the word "lesser" to "necessary." The moment we do, we will glimpse human sustainability for the first time. We saw the power of appreciation during the recent pandemic, where those we once disparaged were coined "essential," and given a free pass to continue working while the rest of us were required to stay at home. They ceased to be simply grocery workers, janitors, and transportation specialists, and we saw what they truly are: the backbone of humanity.

Part 2

Moving Past the Human Injury

Chapter Nine
Short-term Awareness; Long-term Solution

For a fading humanity, the bad news is now overwhelming, and that would explain the totality of a phenomenon called "normalization." I didn't just make it up. It's one of the featured videos on the Noēsis website. These multiple videos describe our mission.

Something pivotal may have become lost in the haze, something completely counterintuitive that you likely did not consider. It's this:

Our ancient ancestors looked for, located, isolated, and quantified bad news, as a matter of survival. For ancient humans, and now for modern humans, survival strategies solve the bad news. The bearers of good news—what value did they bring? Recall that ancient humans plotted and executed survival strategies, all day, every day! The bearers of bad news kept us breathing.

Consider your own experience, every single day. What reader does not rise in the morning and does not quantify his or her concerns? Anxiety motivates every living thing on Earth. Let's consider the following hypothetical and comical conversation between Bill and Steve, who are two friends hiking in the Alaskan wilderness:

Bill: "Hey Steve. There's a Grizzly Bear over there. She has three cubs. She's coming our way. Let's get out of here!"

Steve: "You go on ahead, Bill. I have a pebble in my shoe. I'll hang out here, and I'll catch up with you later."

Steve's lack of anxiety is not possible if he enjoys the privilege of breathing. Humans are motivated by anxiety, or we die. Therefore, bad news is good news. Look at it differently. If bad news is good news, if brutal clarity is a friend to all living things, then why don't we invite our friend in?

With the bad news now viewed differently, many see the vast array of global concerns that humans now face as a new low—but it's not. Recall that human cultures fail with mathematical precision. Every long-gone stationary human culture witnessed their own special version of societal collapse, marked by mounting complexity, fed exclusively by inequality. What we see in the modern day is all too familiar.

But we don't choose clarity. We don't want bad news. We look away. We've normalized dysfunction to a point where it's now achieved cartoon-like status. Normalization of bad news, rather than quantification, is yet another symptom of The Human Injury. The bad news that kept our ancestors alive is now actively avoided, not quantified. Modern humans are actively escaping from what we once feverishly sought.

What would follow escapism? In a word: surrender.

Take a moment to consider the immensity of fully baked-in evolutionary influence upon modern human behavior. We're told that humans are a blank slate at birth. We most certainly are not. When we arrive in our mother's arms, we bring with us the influence of five million years, or 200,000 prior generations, whatever you will. This evolutionary baggage, with its guaranteed misinterpretations, belongs to all of us equally.

NOĒSIS

EVOLUTIONARY BAGGAGE

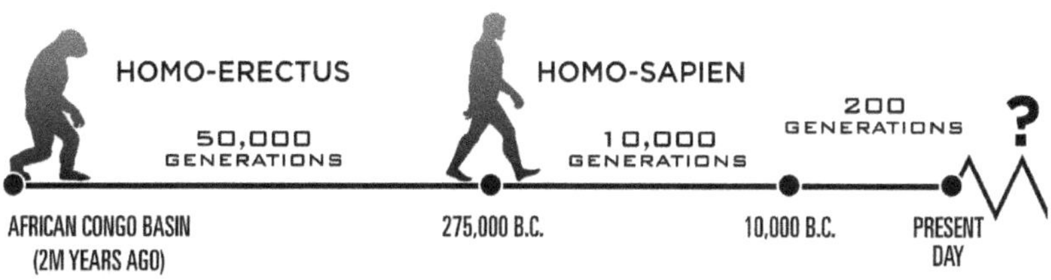

- **50,000 GENERATIONS OF COLLABORATIVE/AFFILIATIVE MIGRATION**
  (100% SUCCESSFUL)
- **200 GENERATIONS OF ACQUISITIVE SELF-INTEREST**
  (UNNATURAL HIERARCHY; 40 FAILED HUMAN EMPIRES)

INARGUABLE CONCLUSIONS:

- MISINTERPRETATION OF ENVIRONMENT
- EVOLUTIONARY HALLUCINATIONS
- MISALIGNED RESPONSE IS GUARANTEED

PRESCRIPTION FOR HUMAN SURVIVAL:

UNDERSTAND EVOLUTIONARY INFLUENCE OR **FAIL**

SCAN THE QR CODE FOR A DETAILED EXPLANATION OF THIS ILLUSTRATION.

A sustainable humanity will fully understand the immensity of evolutionary influence or be swallowed whole by compulsions and impulses so diverse and so many, they're impossible to quantify let alone defeat.

Modern humans have yet to remotely acknowledge our long-ago expelled, exponentially darker alter egos. In literal terms, then, we have no earthly idea who we are. These evolutionary impulses and compulsions never packed their bags, even as their ancient enemies ceased to exist. These baser compulsions don't think. Their job, in fact, is not to think. These ancient ghosts, these agents of brutality, these mindless thugs still alive inside of us rule over modern humanity. Achieving escape velocity from this serial killer is impossible, in lieu of repetitive and unceasing educational initiatives, using evolutionary data we now have.

It's true that humans are built to locate and manage bad news—to a point. The bad news for modern humans has piled up considerably. Its immensity exceeds the human capacity to manage it. As we run for the exits, we bear the weight of fifty pounds of collective insult in a 20-pound sack. Escapist behavior should be regarded with great empathy.

SPECIES SELF-LOATHING

We guilt-ridden modern human beings, as a collective species, do not like ourselves. But how could we? How can a sentient being, one that knows not how it got here, or how it's leaving, be blamed for aimlessly searching for itself in the dark when it is gravely injured?

But it gets worse. This same being, already put-upon, is now witness to its own degradation. What do we expect of ourselves? What will it take to move us from the self-loathing we don't deserve, to viewing ourselves with great empathy, an arrival point that we deserve?

By example, humans feel great empathy for a whale that's entangled itself in a fishing net. How are we different than the whale? We're a helpless and frightened species, entangled in our own net of evolutionary baggage. The weight of six million years of evolution—100,000 generations—is an onion to peel, not a scourge to loathe. When self-loathing stops, self-preservation starts.

Blame is one of the pieces of evolutionary baggage that we're all carrying around. It rings cash registers today, and it will also ring the island bell tomorrow, to announce the passing of the most remarkable, rare, and helpless species in the Milky Way galaxy. We are all blameless for The Human Injury. None of us was alive when we randomly started agriculture to feed our growing populations. How can we blame ourselves?

We face low odds. I never said it was going to be easy. I never said it would be comfortable. Let's take a shot at beating mathematical odds.

Fighting Fire With Fire

Here's some lighter fare. Let's pretend that we are standing at the edge of a huge forest fire. We're looking at this raging conflagration, completely out of control. I turn to you and say, "What are we going to do about this?"

You might say, "What can I do? I'm not a firefighter. I have no tools to use to stop the flames. I can't do anything." That would be normal.

We take out our cell phones, and we call the fire department. They arrive with their heavy equipment, but not in the blink of an eye! Look, no one can stand at the edge of a fire called perpetual conflict and wave a magic wand to extinguish it.

This analogy is imperfect because we have the infrastructure to fight fires. What about the infrastructure to dissolve eternal chaos? That's something we'll need to design and build. The design phase looms large, so now might be a good time.

Imagine now that we're standing at the edge of the same forest fire, but there's no such thing as fire engines and hoses. The only thing we know how to do is to fight fire with fire. Do you see where I'm going here? Nowhere, and fast.

Blame is a similar creature. We are standing at the precipice of human extinction, ready to fight it with the gasoline of failed models that started it. And we're blaming. There's nobody to blame, for God's sake. There are no people behind this, no conspiracies, no master plan. It's a *thing*. It's the only path we were shown. And its origin is now well known.

The Human Injury happened ten thousand years ago, when a masterpiece of collaboration and affiliation called *Homo sapiens* became self-interested.

I can't prescribe a pill or promote any piece of legislation that will solve The Human Condition—but we can redefine it! The most vital step is understanding what is happening to us. Only then can we attempt to come about. I'm referring to severe, imbedded, societal problems that took ten thousand years and fifty dead empires to develop.

We're crossing the threshold. We're staring into the void. Nothing but a complete restructuring of our relationships with ourselves and each other will be enough to cheat The Great Filter to emerge as sustainable.

Chapter Ten
Drivers of Perpetual Conflict

Human conflict is rooted exclusively in evolutionary paranoia, the egregious misinterpretation of our environments. Another log in the fire saw the formation of unnatural hierarchies within stationary/acquisitive cultures. In its early stages, acquisition seemed like a positive development. If a person could cultivate food and store it, shortages and famine would be less likely.

Commitment to Extinction began with our transition from a life of service to others to a life of stationary acquisitive behavior. Ground zero for The Human Injury is the word "stationary." The DNA blueprint for responsible stationary behavior is missing. That's human biology, not human scourge.

Could it be, then, that every single conflict to ever occur on Planet Earth started as an evolutionary misinterpretation, and escalated from there? It's not merely probable, it's obvious! There have been trillions of conflicts over the last 10,000 years, both micro and macro. These conflicts never needed to occur, had we deduced evolutionary influence. We couldn't—but now we can!

No conflict that happens from this day forward will ever need to, not with these evolutionary motivators known. But don't tell anyone—not yet. If we tell global financial markets that human conflict can be phased out, we won't be congratulated. No revenue stream tied into conflict will ever surrender itself, nor will they be asked to do so! If we're short-sighted enough to threaten these failed models before they're ready to negotiate, we'll disappear faster than a scalded cat. Conflict feeds our children; it pays our bills; it makes us rich; it makes us die horrifying deaths. It builds and maintains our war machines. Like all addicts, therefore, these models of destruction will protect their heroin with fury. Over the next century, these models of abject destruction will deduce their own uselessness, or our species will be dead faster than s*** through a goose. For those who find my salty language to be offensive, I find your inattention to be deadly.

THE HISTORY OF PERPETUAL CONFLICT

In Greek philosophy, *Noēsis* is the exercise of reason. It denotes the highest attainable level of intellect, using all available data. I named my organization Noēsis after an ancient belief system that sought data that was never once available to it. Consider the civilizations below, as well.

1. Mesopotamians invented the wheel. They explored the mystery of time. They pioneered bricks for building. They did not deduce evolution. They succumbed to evolutionary hallucinations, misinterpreting everything they saw. It placed them into conflict and kept them there.
2. Ancient Greece had Socrates; Socrates did not deduce evolution. Their will to survive was notable, but they did not deduce evolution.
3. The Ottoman Empire took great leaps in medical technology, using forceps in childbirth and catheters and scalpels in surgery. They had advanced mining technology. They did not deduce evolution. Perpetual conflict, with its assured societal collapse, was their only destination.

The constant for all these civilizations above was a loss of escape velocity authored by mathematically preordained failure. Consider that no automobile without brakes can stop its inertia. Neither did any past civilization, missing the human flight manual called evolution deduced. Again, they did not succumb to their faults. They succumbed to the undetected and untreated wounds guaranteed by The Human Injury.

MONETIZING CONFLICT

We see the common enemy, the very mortality of our species, just up ahead. We understand the totality of once vital, but now useless evolutionary mechanisms that place modern humanity into unnatural, ceaseless, and mindless conflict. And yet, with all we can see, with all the data available to us, we're paralyzed, due to a reality that I earlier expressed. You recall it. There's no money in the fix, only in the fissure.

Without the heroin of conflict to ring cash registers, we have a whole new set of problems. Perpetual conflict is fully monetized. That's always been true, but this time globally. What has always destroyed us, also sustains us. We have no bridge to sustainability. Our freefall has no parachute.

• • •

When the COVID virus sneaked up on us, it began as a sound bit, an interesting one, but not one that commanded attention, let alone anxiety. After that, we were out of the loop for a few days, busy with work. The news soon transcended multiple media platforms and then crept into our conversations. Those conversations were then everywhere we went. As Americans, we saw a concern that would be isolated. It would never reach our soil. Looking back, our naivete was notable. We can now fly to any destination in the world in 24 hours. So too can a pathogen. Still, we weren't ready to let it in, the idea that COVID had now reached the front page, and days later, it reached the top of the front page.

The financial markets came next, collapsing with a fury not seen since The Great Depression. On February 20, 2020, vigilance crossed over to panic. The virus had now reached New York City. A week prior to that, the Dow 30 Big Board was poised to break through 30,000. By early March, a whopping 40% sell-off had by then wiped out trillions of dollars in global wealth.

Why would I describe that which we already know so well? We all noticed the collapse, right? But I noticed something else, namely, how little it took to evoke a worldwide panic. What else we didn't notice was that the prime directive for all human beings on Earth, one that enabled our ancient ancestors to migrate across the globe, was completely absent. What took its place? "Collaborate or die" became "protect myself or die!" Covid and its self-interested arrival points signaled humankind's full inability to survive itself. A self-interested humanity is dead on its feet. The observation is harsh; it cannot be defeated. The battle is afoot.

Let's return now to the mission.

The misinterpretation of our modern environment (i.e. evolutionary hallucinations) is by itself destroying the human species. Humanity will never defeat this constant unless it fully owns and understands it. Otherwise, human failure is imminent. Evolutionary data that was never available can slowly shift human thought. Should humanity choose to know what is now knowable, and should it come to value itself by that decision, we will only then preserve ourselves. Achieving sustainability assumes that we have not already mathematically committed to our demise.

Have we committed? I'm certain that we have not. Are we fence-walking between sustainability and oblivion? We're headed out the cosmic door, and you know it. I ask again. What are we prepared to do?

• • •

Show me an extinct species, and I'll show you an unlucky species. We modern humans must henceforth view our reality in an entirely forensic fashion. We are forensic in nature. We're built to explore. We're capable of greatness. We know everything about everything, and almost nothing about ourselves.

Humans are machines of extrapolation. We examine objects and make them better. We do that because of two words no one thinks of. "Opposable thumb." (Tape your thumbs to the side of your hands, and you're in for a long day, friend). We're built to evolve, even as we devolve to commitment. Human degradation now has the rarest creature in the galaxy awaiting its fate, while our physical bodies have tools meant to escape our fate.

Perpetual conflict is all around us, again. Our children leave our loving arms to engage with a world that seeks shelter from its many storms. Our children wonder, *what is this awful place?* We told them they could run. They learn quickly how to duck. We've been at one another's throats for ten thousand years. We were stumbling then, and we're stumbling now over the residue of ancient compulsions that still burn brightly inside of us.

Chapter Eleven
Releasing Blame

I have several friends who are lawyers. Lawyers show up when people are at their worst. Our justice system is doing the best it can with a caseload that exceeds its capacity by light years. Into this thankless reality comes an army of attorneys. The cheap shots we take at attorneys fail to consider that these professionals are tasked with the impossible: to successfully lie.

Counselor, I'm about to do you a big favor! Let's change the words of a famous lawyer joke. Let's surgically remove the nonsense and deftly implant the truth. We start with the nonsense:

> *How can you tell if lawyers are lying?*
> *Their lips are moving.*

That's complete nonsense. Here's what's really going on:

> *How can you tell if lawyers are lying?*
> *Their lips are moving, to tell* ***OUR*** *lies, not theirs.*

I'm promoting attorneys all the way up from the born liars we call them, to the unwilling delivery conduits for the excuses we make for acting on our baser compulsions, and then denying the existence of these compulsions. It's a neat trick, one that will kill a clueless humanity. I've seen enough!

A sentient human species that keeps an average of fifteen major secrets per person is late for the cosmic trash chute. Humans are compelled to lie, or we risk being shamed by other liars who stayed with their lies. The Human Injury created, monetized, and now demands our endless lies.

A few of my lawyer friends are delightful people. One of the folks involved with Noēsis is a top-notch attorney. He's totally unimpressed with himself. He'll know who he is when he reads this. He'll recall this story. Over a beer one evening, I asked him, "What is the basic foundation of the law?"

He didn't regale me with the rights of citizens to equal representation and the bedrock of the legal system. He knew better.

"Where are you going with this?" he asked. He's intuitive that way.

I then risked the following. "Tomorrow, the clients that you represent will each ask you to do one of two things. You will spend all day assessing blame or rejecting blame. Is that fair?"

Now, this gentleman is no pushover. He thought about it for a minute, and finally said, "Come to think of it, you're exactly right. With some alternatives, you're correct. That's exactly what I'm going to do."

He'd never thought about his job in that way. There are television shows that are dedicated to just that kind of insidious joy. Not only have we set up our entire system to assign blame, but we enjoy watching the spectacle play out. This model is as childish as it is essential. We're acting like children, while also having no other choice but to act like children.

THE PAYOUT OF BLAME

"(We've) paid our dues; Time after time.
(We've) done (our) sentence but committed no crime."
- Queen, "We are the Champions"

Watching courtroom dramas is every bit like rubbernecking a highway accident. You slow down to contemplate what life must be like for the unfortunate soul whose years were cut short or altered forever and thank your lucky stars you are safe.

Our judicial system pretends to take circumstances into account. How can it regard the circumstances when the chief objective is to assign blame?

But it's not that easy, is it? Remember all those ringing cash registers? Retirement of blame/blame rejection would silence our employer. If we throw a switch to stop it, our financial markets fall to $0.00. It will take a century to unwind it. We'll back out of this mess over decades and decades of time, or no effort at sustainability should even be considered. We hired our executioner, but they won't take our pink slip well. They'll need a severance package—a long one. Should anyone ask our grouchy employer to give up even a plugged nickel on the way out, a sweatshirt with a bull's-eye on it is my fashion suggestion.

A message to big business, particularly those who depend upon conflict to thrive. This is not a walk of shame! This is an invitation to ply your wares on sustainable streets, while giving up nothing in the short term. We awaken in the modern day, to the most hopeful and wonderful of surprises. It looks like this:

"There's always a chance,
a tiny spark remains.
And sparks turn into flames...."

- "I'll Be Around" by the Spinners

There's an old cliché in the human lexicon: we're going down the tubes!! I take it a step further and identify the main tubes that modern humans use to achieve perpetual conflict. We can now see these conduits and thereby know where to focus sustainability efforts.

NOĒSIS

BLAME/BLAME REJECTION

"UNNATURAL BEHAVIOR CANNOT PERSIST"

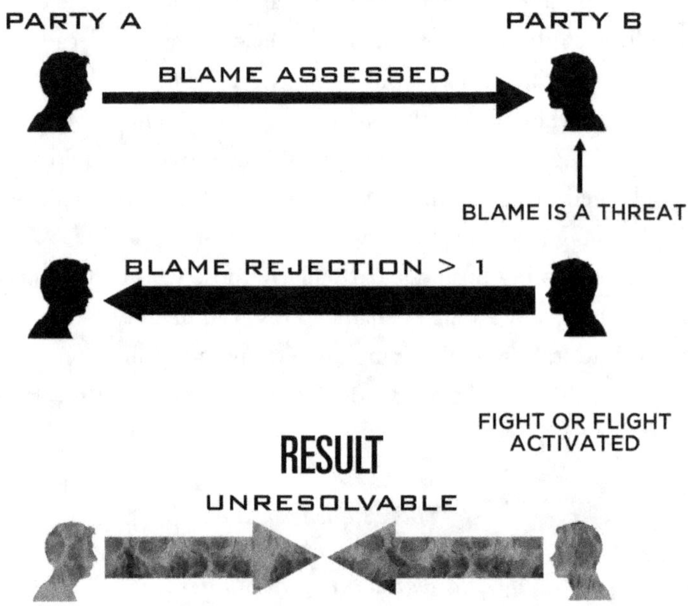

INARGUABLE CONCLUSION:
THE ABOVE DYNAMIC = Ø

PRESCRIPTION FOR HUMAN SURVIVAL:

REDIRECT OR FAIL

SCAN THE QR CODE FOR A DETAILED EXPLANATION OF THIS ILLUSTRATION.

Blame/Blame rejection goes directly to perpetual conflict. It does not pass go; it does not collect $200.

Make no mistake, blame is an immediate threat to the human mind. We will attack and kill the mere concept of blame precisely because it's a threat. By this simplest of observations, see everything in front of you differently now. Nothing—and I stress this—we see in the modern day is what we think it is. When the very business model of a beleaguered humanity places us on the edge, it's a one-way ticket to over the edge. Again, behold the brutal nature of mathematics and its first cousin deductive reasoning. The arrival point of these models is not open to debate of any kind. The assured failures of blame/blame rejection, deployed trillions of times per day, are so many metaphorical lemmings heading to oblivion.

Eight billion modern human beings, now emotionally exhausted, were placed at one another's throats by a guaranteed failure, and we're now kept there. What's changed? It's this. We see it! That is the first step for the salvation of these trillions of lemmings. Seeing the target, seeing this mathematically assured failure—seeing this is necessary if we ever wish to alter its inertia. It's on our radar, and we can now set a course to intercept what it is.

We're not finished here. Remember, this is not a dark undertaking. This is not a walk of shame. Why can't human salvation be fun?

Here's your assignment:

- Pick out any ten random people. Man, woman, black, white, red, yellow—humans are humans. Pick out ten.
- Blame them for something. Perhaps they litter. Perhaps they hold up traffic. Tell them to right their wrong. Say this: *"Hey, you just littered. Pick that up."*
- Get back to me. Tell me how it went.

Let me save you some time. Here's how it went:

1. Your blame was rejected straightaway, with even stronger force than your blame.

2. Guilt and innocence are irrelevant. The human brain won't bargain with blame.
3. The immediate escalation from peaceful to chaotic stunned you.
4. Your blame was pulverized. It was not acknowledged. Any questions?

We'll get to the concept of misaligned force. If you blamed someone today, you already have a nice head start. The escalation you witnessed, the one that might get you killed, is suffering from evolutionary hallucinations. There are no pills for the heavily misaligned force that descends from evolutionary hallucinations. The threats we scanned for long ago ceased to exist. In fact, the ancient predators we seek to this day, their bones are powder by now.

We're locked into this cycle like a broken washing machine. Blame is insidious and unsustainable. Any behaviors that are precisely predictable (with some outliers) are socio-mathematical constants. We want that ownership that our blame demands. We need that ownership, if even a tiny bit. That might return us to the collaborative method of interaction that burns inside of us. Alas, the very solution we've deployed pushed us farther away from the communion we seek. Blame doesn't make us safer. It doesn't provide the ownership we crave. We're farther away than ever.

A toxic brew indeed. But not so fast! Cup your hand to your ear. What you hear in the distance is our cash registers ringing. What fails, also sells. Blame giveth and taketh away. Matter and anti-matter annihilate. How can we hope to survive an action/reaction model that can't survive itself?

I'm not open to any interpretation that views an innocent humanity as the author of its own demise. We've located and isolated the evolutionary crime scene called The Human Injury, whose precipitating event is visible.

NOĒSIS

NON-CONCESSION & STUBBORN
THE ANCIENT MECHANISMS

Non-concession and stubborn behavior share the same evolutionry biological mechanism as illustrated below.

1. "Party A" (below) states an opinion
2. "Party B" does not concur.
3. "Party A" perceives a threat. The threat is imagined.
4. "Party A" responds biologically. Cortosol increases signals. "High Alert"

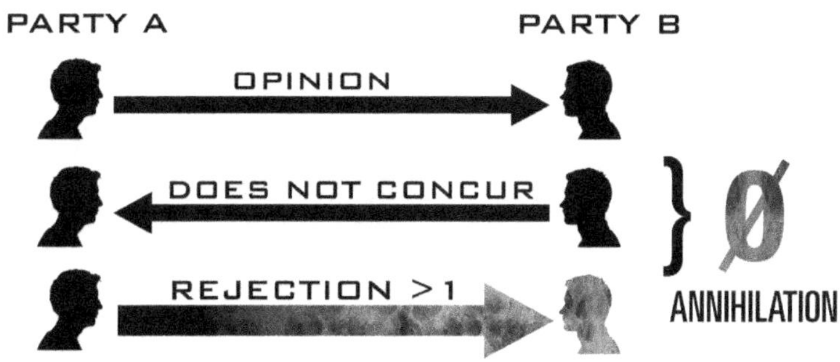

CONCLUSION

HUMAN CONFLICT IS NOT RESOLVABLE. SUSTAINABILITY = Ø

SCAN THE QR CODE FOR A DETAILED EXPLANATION OF THIS ILLUSTRATION.

Chapter Twelve
Clarity

In the 21st century, modern marvels like the internet and smartphones have turned us into islands. That's why I'm so happy you're here, reading these words. The clarity I impart is designed to focus and engage you, not frighten you. If I can get you into that space, where you sense no agenda, then we will find we share a common concern: that modern mankind is wildly off course. We'll also agree that "wildly off course" is suddenly a reasonable assessment, not an alarmist one. Escape has never solved anything. In fact, escape is unnatural. All of us notice and quantify our concerns daily. Our behavior is not merely notable, but vital. We can thus agree:

Clarity is the best friend of all living things. Without clarity, we die.

The clarity we have can quantify evolutionary influence on our behaviors. We can know the ancient triggers for all modern behavior. We can examine the crossover period of ten-fifteen thousand years ago, when larger populations developed and when creatures born to run became stationary layabouts. We can view it as an injury, never a choice. We don't need to guess the right answers anymore. We can know them. The test that we failed time after time is one we could never study for. All of that ends now. We can restructure our failing models over decades and decades. We can abandon the illusion of greater beings and we can assuage the resentment of lesser beings. We can become necessary beings, the ones we were born to be. We can slow and arrest human degradation. I'm asking an entire species to abandon its false bravado, to stop clinging to the edge of its cliff. I'm asking humanity to engage in a bid to preserve the most remarkable and rare species in the Milky Way galaxy. I'm asking that we band together as one species of like and equal beings, to free ourselves from our prison of evolutionary hallucinations, to understand and to stop misinterpreting modern threats, to learn and expose the evolutionary triggers for all human behavior, to reintegrate with our darker compulsions, not to expel them, but to understand them.

I'm asking that humanity embark on an endeavor of self-discovery through the conduit of education, to learn who and where we are, how we arrived at this awful place and only then to plot where we need to go.

Inaction will not be considered or an inactive humankind will perish. We will not stumble into sustainability. It will look like this:

- What we have become must reintegrate with what we were for 50,000 generations. Acquisitive and collaborative must seek communion.
- Should we not, the two are doomed to annihilate until the earth falls silent.

Racism's warped conclusions demand a global educational initiative designed to explain the oneness of human origin. Perhaps a future exercise, where every human on Earth wears the same T-shirt emblazoned with the word ONENESS on the same day, would help what's so easy along.

The obstinance that modern humans wear like badges of honor channels an ancient and unresolved compulsion. The ancient enemies that human obstinance holds at bay are long dead. Humans are at one another's throats, protecting themselves from dangers that no longer exist. A fight requires a threat. Where is the threat?

Across every religion in the world, there is a singular driving force that feeds every decision and action: compassion and concern for others.

Conventional wisdom, that any individual can succeed on their own, is tragically false. In fact, many human beings are dealt a bad hand by the birth lottery. Being born is a statistical miracle. Every living thing in the universe is a masterpiece, painted by an artist so utterly elusive that we struggle even to name it.

Chapter Thirteen
Helping Others

Noēsis will assist in teaching the entire human race the glory of helping what is helpless. I won't merely promise it. I'll do better. I'll prove it right now.

We've all been in trouble. Whoever says they have not survived personal calamity is yet another casualty of The Human Injury. [The stiff upper lip affixed to the modern human being is a ruse. I stress, it's a lie and a bad one. Once we abandon the nonsense of conventional wisdom, we'll discover the new boss called counterintuitive thinking.]

Let's deploy counterintuitive thought to bring a dead soul right back to glowing life. This is how we help what is helpless—watch how it works.

How to Help Another Human Being

A Field Guide

If a friend reports being in crisis, counterintuition must be brought to bear. Only counterintuitive thought can rescue anyone caught up in a Darwinian nightmare of opportunistic abuse.

Any crisis reported to you by another is already bigger than you could ever know. Either pride or shame has kept this person from surfacing. They surface now only because the dam broke. This crisis can no longer be managed. Your friend has a survival issue. Do not underestimate its intensity. Do not impose your will upon someone's crisis, especially a valued friend. Instead, deploy the following steps to bring meaningful relief, to truly stabilize the situation:

STEP 1:

Your friend has needs that cannot be met until fully quantified. You can't guess their problem. Do not default to your instincts. Instead, promise your loyalty. That will summon the most honest description of these concerns. Quell their fear and encourage 100% disclosure of the details. Make them feel safe. Only then can you systematically define an invisible maelstrom, starting from the ground up. There are always immediate dangers. Start with these.

STEP 2:

Ask them, "What will it take right this very minute to stop the bleeding?" Your friend has fallen for reasons that you cannot see. There are basic needs, like food or an unpaid bill, that risk the loss of power or water. Assess immediate needs and solve them. Then place them back onto the ledge.

STEP 3:

Nice job. You've placed your friend back on the ledge. But leaving them there on that ledge and walking away risks them falling back off. You're not finished here. Assume that your friend feels crushing fatigue caused by a battle as long as it is lonely. What will it take to move your friend completely out of danger? Quantify it! Get him or her off that ledge. You must carry your friend out of a house that's burning down around them — the one you can't see. Don't carry them partly out. Carry them out completely. Whatever that looks like, do it.

STEP 4:

With steps one through three completed, you've secured the situation. And even better, you have not judged your friend in any way. Perhaps only a single day has passed. You've already done what most people never think to do. You quantified, located, and stamped out all emergencies. Only now can we pose our next question: *How did your friend get to this awful place?* Is this a drug or alcohol concern? Has mental illness gone untreated because of shame or stigma? Has a financial crisis crippled your friend?

Relieve the burden for a few days. Gift to your friend their first good night's sleep in months.

STEP 5:

You've rescued your friend from all immediate dangers. You've assessed the cause of the crisis. Now, what will it take to guide them to sustainability? Illness or addiction may be the underlying cause. While conducting your investigation, please, let your friend know that you care deeply. You're looking at a silent and intense battle, whose symptoms might have been spotted weeks ago. But you couldn't help what you didn't know. Don't linger there.

STEP 6:

With illness or addiction assessed, with a remedial plan in progress, and with food in their refrigerator, only now can clear and long-term thinking take place. And strangely, only a few days have passed since these problems were made known to you! That makes perfect sense, doesn't it? Explosions are intensely violent at their beginning. That will give you additional insight later into how truly awful this battle was.

Expect money to enter the picture again here. If you are inclined to assist, please trust a friend to help themselves. Don't micromanage financial needs! This is your friend, not a child. They could use some confidence right now. Your friend is injured, frightened and ashamed. You cannot set terms, for the most basic reason. This isn't a negotiation. What terms exist between friends? Set a limit if needed but keep that to yourself. Your friend will later strive to live up to your trust. Isn't that how you became friends in the first place?

STEP 7:

It's finally time to share this crisis with mutual friends but maintain a disposition of unwavering empathy. Never devolve into the betrayal of gossip. Identify and gently expel all disinterested parties. Don't be surprised if the first person you expected to help is the first person to turn their back. We reveal ourselves in these moments when decency requires not only our words but our actions. Whatever judgment occurs to you, tuck it in! It's not up to you to judge these individuals. Perhaps their own, unseen hardships have intruded. Their choice of abandonment may one day haunt them. Move on.

Compassionate friends will wish to help. Encourage them to reach out to your friend and to be plainly honest. Talk to them as you always have, in the language you've always shared. "Look at me, friend. We've laughed together for years. Now life intrudes. Talk to me, and let it go." When an injured friend feels safe in their humanity, they will express themselves openly. They will recount the nature — and the trauma — of their experience. Their descriptions are cathartic. They may cry intensely from relief, due to a new reality, one of hope that you gifted to them. Congratulations. You're the author of a miracle.

If judgment, abandonment, gossip, intolerance, avoidance, and blame replaced those steps, your injured friend would circle a drain, and disappear, helped along by our primitive tendency to diminish the weak. Modern humans think that we're staring up a mountain to achieve sustainability. We most certainly are not. If acting as we're intended to act, to help what is helpless, this is no mountain. This is a walk in the park. Counter-intuitive!

We're at our best when we're needed. Lack of compassion deadens the soul. Your words and actions are a great gift, not only to your afflicted friend, but to you as well. "A rising tide lifts all boats." Mother Nature may have just changed her mind. We build from here.

PART 3

A SUSTAINABLE FUTURE

Chapter Fourteen
Where Do We Go From Here?

The dawn of acquisition was our greatest undoing, though we didn't recognize it as such. We didn't set out to knock over the thousands of dominoes that have led to our current level of crisis. No, the adaptation of acquisition was an outgrowth of several natural tendencies which I will now explain.

The crossover from collaboration to acquisition by itself selected the human species for its eventual demise at its own hand. Because I say so? Hardly. I earlier explained, and I now repeat, that nature will not tolerate unnatural behavior. The only workable economic model available to a sentient and free being is capitalism. In a paradox so exquisite in nature, therefore, the only model that makes sentient beings free also dismantles and destroys stationary cultures on planet Earth. There are no exceptions. For any reader who disagrees, I'll wait here. Please bring me an example of unchecked financial meritocracy that ever survived. (Place the Jeopardy jingle here. Let it play into eternity. It will never be disturbed.)

The Economy

The only workable economic model for humans is capitalism, but the decisions we make in a sustainable capitalistic future must replace the word "lesser" with the word "necessary." If even one diminished human being is observed, the collapse of the system he or she inhabits is a mathematical assurance.

Recall what we're doing here. We're problem-solving for human sustainability. We're identifying socio-mathematical patterns that descend from a viewable event, the human crossover from collaborative to acquisitive behavior.

We have no choice but to understand our evolutionary journey on Earth. These impulses crowd out the thinking mind. Our past reality, our present reality, and our future aspirations are listed here:

1. **Collaborate! Or Die!** (Our Migratory Past)
2. **Acquire! Or Die!** (Present Day—Not Survivable)
3. **Understand The Human Injury, and Live!**

Our migratory past is over. Our stationary defaults have uniformly failed us. We're failing again. Understanding the crossover event of 10,000 years ago - from Collaborative experts to Acquisitive killers - will gift to humanity its first chance to survive.

If #3 is ignored, we're DEAD ON OUR FEET. Spare me any nonsense that rationalizes perpetual conflict as survivable

Is there a solution? As it now stands, no. As it must stand, yes. Or we will die. Without the recognition and ownership of socio-mathematical constant ignited by a crossover from who we are (collaborative) to what we are (broken), we're all finished here. This is REALITY. The paperwork for the human tombstone is on our desk. Will we sign it, or will we reject it?

HOPE FOR THE FUTURE

There is plenty of hope, or I wouldn't even start this book, let alone finish it. But there's no hope if we look away! If another argument can't put me out of business, and if the facts are ignored, no amount of hope or prayer will cancel modern humanity's path to oblivion.

The deepest of human longings, to collaborate and cooperate, to assist and to be assisted, to help what is helpless, to appreciate, to belong — all of these are at odds with models of unchecked meritocracy, unnatural hierarchies. The self-interest of acquisition collided head-on with these deepest of longings. Matter and antimatter annihilate. This is modern man, acting unnaturally, smugly ignoring a watchful and indifferent universe.

A sustainable humanity, fully aware of itself, will deeply understand and only then avoid these excesses, or we will simply perish. We knock ourselves out of the ring every time. We will rely on one another. Or we will die. Collaboration is not a choice. This is evolutionary science, people.

Bad news kept our evolving ancestors alive! Let's allow it to keep us alive now. As scary as extinction may be, it's nowhere near as terrifying as commitment would be.

The Blueprint

What now? What is realistically possible for a sentient species that quantifies its prison, to then attempt an escape into sustainability? Only this:

1. Humans will access data now available to them. We will know the ancient triggers for all modern behavior or be swallowed whole by compulsions that activate instantaneously, that grossly misinterpret everything they see, that override the thinking human mind. We will only then glimpse sustainability.
2. We will surrender in full to the formidable power of ancient relics of evolution. They are our puppeteers; we are their marionettes. We will bargain with them, not they with us.
3. They do not know that their ancient enemies are long dead. Their actions are grossly misaligned. A process of proper alignment will unfold over decades. It guarantees nothing whatsoever, other than increasing chances of human survival to above zero.
4. No process that includes lesser humans will survive, regardless of methodology. Humans will view all others as like and equal, or no attempt to find sustainability will bear fruit.
5. The currently ruinous socio-mathematical inertia will be altered by the above process, unfolding over decades, through the conduit of slow-drip education. Human thought will undergo a seismic shift, or our sentient species will join all others before it, in accordance with The Great Filter. In other words, we will die.
6. The inability of sentient intelligence to escape technological childhood will claim its latest victim. Should humans achieve it, what is eminently normal will be called a failure.

Our opponent is The Human Injury. What weapons can we bring to bear upon a random evolutionary crossover event? We just deployed one, and very few readers even noticed. It's called knowledge. We now know something we never did. Assessment is always the first step.

Only a being that understands itself in detail — as we can now — can plot its next point of arrival. Only a vehicle of slow-drip education will deliver our like and equal species safely into the future. When we examine humanity's evolutionary past, we easily deduce that modern-day behavior springs from a past reality, one whose enemies and predators are long dead.

I suggest that we deduce the obvious mighty fast, too, before we reach an event horizon, the mortality of the most rare and remarkable species in the galaxy. Pardon the weight of these descriptions, but recess is over. Water can only circle a drain so long before it disappears forever. That awful place is vaguely visible to all humans on earth, the point from which we cannot return. Commitment!

What we need to know is now so knowable! Man has always created educational tools. These new tools of education can be standard. Humans all come from the same place. Without these tools of education, humanity will continue to degrade. Math says so. Accept my apologies for an unwillingness to soft-pedal the very nose on our face.

We need a change of direction here. Try this one on for size:

History shows that humans unite over common enemies. My God, is the mortality of the human species not common enough? We're about to find out.

UNITING TO SAVE OURSELVES

Before we proceed, we'll do well to define the categories under which humanity cannot unite. I think the reader will agree here:

- Can we unite politically? Of course not. Why would we attempt the impossible?
- Can we unite religiously? In a pig's eye. We will not pursue what cannot occur.
- Can we unite culturally? Nope. We're not ready — not even close.
- Can we unite under the banner of species' preservation? Yes. It's doable!

NOĒSIS

IS HUMANITY SUSTAINABLE?

CURRENT TRAJECTORY

- 40 STATIONARY CULTURES ARE GONE
- CURRENT GLOBAL CULTURE FAILING
  (PERPETUAL CONFLICT MONETIZED)
- ILLUSION OF CIVILITY
- ILLUSION OF COMMUNITY
  (PRIVACY IS UNNATURAL)
- UNNATURAL HIERARCHY
- XENOPHOBIA
- NORMALIZATION
  (THE ARRIVAL POINT OF HELPLESSNESS)
- ESCAPISM
  (PLEASURE SEEKING/PAIN AVOIDANCE)
- EVOLUTIONARY INFLUENCE
  (NOT QUANTIFIED)
- SPECIES SELF-LOATHING
  (ESCAPIST BEHAVIOR)

REDIRECT TRAJECTORY

IMMEDIATE GOAL

GLOBAL CONSENSUS
TREND AWAY FROM FAILURE
EXISTING ECONOMIC MODELS UNDISTURBED!!

MID-TERM GOAL

LESSER BEINGS BECOMES NECESSARY
EVOLUTIONARY INFLUENCE QUANTIFIED
ONENESS OF ALL HUMANITY REALIZED
SELF-LOATHING TREKS TO EMPATHY

LONG-TERM GOAL

SPECIES-PRESERVATION BEHAVIOR
COMITTMENT TO EXTINCTION AVOIDED?

SCAN THE QR CODE FOR A DETAILED EXPLANATION OF THIS ILLUSTRATION.

A put-upon humanity can, in fact, forestall human extinction. None of what I describe is remotely necessary. I've called this current human trajectory an inertia, not a choice. An approaching storm is one that we can dodge. The evolutionary data is there! A mathematically certain arrival seeks an altering force. I created Noēsis to provide it.

Importantly, the word commitment does not denote a purposeful decision. Commitment denotes the point past which failure is not merely imminent, but mathematically irreversible. The human family is fence-walking between sustainability and commitment right now. As much as you don't want to hear it, that's also how much you know it. All the historical ingredients for the failure of yet another stationary human culture, courtesy of The Human Injury, are securely in place. The human family must now summon the courage to quantify its reality.

Should humans not recognize, admit to, and own the following cognitive distortions, make no mistake. They'll swallow us whole. A sustainable human species would identify and expel its delusions. Stripped bare, we would only then meet ourselves for the first time. The bubbles of fantasy that I pop, all over the globe, will be audible from where I live. Truth is always extinguished first and congratulated later. When we survive these delusions, thank me then.

THE ILLUSION OF CIVILITY

Humans are plagued by cognitive distortions. They descend chiefly from a time when we finally imposed laws upon ourselves. The laws that we imposed upon our behaviors made humans safe to prosper, relative to earlier, more dangerous times. The ability to walk up to any given stranger and kill them was once our reality.

I can only speak for myself, but the laws that stop a perfect stranger from killing me in the street have done wonders for my productivity. Those laws protect us, but alas, there was major blowback. Those laws fully demonized and denied the very existence of a decidedly darker side of humanity. Our laws, and especially our religious and philosophical disciplines, are no match for evolutionary compulsions and impulses that activate instantaneously.

These banished compulsions are now spoken of in the third person. I'm describing a disintegrated being. The human attempt to defeat its biological longings is a failure. Any attempt at human sustainability will collapse under the weight of what we will not view. We're only as good as the secrets we keep, and evidence suggests that those secrets are many. Our laws keep us safe from others. Our illusion of civility will kill is. Only a human species that knows the calamities it's capable of creating can plot workable strategies to redirect, in the name of its very survival. Will the addict take inventory, or will the addict self-destruct? There is no middle ground.

Let's identify, in no uncertain terms, the illusions under which we labor. Should we not abandon these distortions, I say again, we will die.

THE ILLUSION OF COMMUNITY

Modern humans refer to our social models as being part of a community. An inconvenient reality must intrude here. What do we truly know about our neighbors? Nothing. Do we invite the community into our lives, or do we retreat into privacy? The latter reality, privacy, is unnatural to humans. We are affiliative beings! When collaborative and affiliative beings retreat into complete privacy, and reject any attempt to invade their territory, that is The Human Injury acting out right in front of us. Neighbors? What neighbors? These are more correctly, "the other," something to be feared. Why do we retreat into privacy? Elementary. If we open our broken hearts to the community, they'll be further shamed by the betrayal of gossip.

We shame our own! My God, what do we expect when our community beckons us to come out and we're then shamed back inside?

No delusion is sustainable. Modern humanity huddles in shame and tells itself that all this will be fine. It will be what we make it but only when we strip away the lies we tell ourselves. We've got so much work to do. It begins with stark and unblinking truth. In an ending to this observation that I dread writing, we, the lonely people are saddled with something less obvious than the tragedy of intense privacy. These models that shame us into privacy also pay our bills. The medications we then smother our wounds in are the most heavily taxed. We educate our children off the backs of the brokenhearted. All of this will cause the normal person to turn away. I understand. I risk it all by telling you this, but the human species that I so greatly love is at risk if I don't. I really don't know where to take it, so I brought it here. Lucky you!

Human beings are acting in an unnatural fashion. By this metric, we have in fact devolved. Natural selection has no reverse gear. Human beings are selecting themselves for their own demise. Viewable socio-mathematical constants drive that arrival. Irreversible commitment is just up ahead for humanity, should we not access and utilize evolutionary data that is now available, to determine the evolutionary triggers for all modern human behavior.

ACQUISITIVE MERITOCRACY

A sustainable humanity must, absolutely must properly align its choices. We're finished with migrating. There are no places to migrate to. At risk of oversimplification, if we agree that every human being is equal and necessary, what we call lesser beings would slowly meander away from resentment and would thereby be less prone to the plot vengeance. Resentment's instant default to vengeance would dramatically decrease. Behaviors of insurgency that descend from vengeful (read: broken) humans would ebb, in a future that recognized all humans as uniformly equal.

Existing models of acquisition are here to stay, but their ruinous defaults can be adjusted. If the radio in our car is too loud, do we risk our hearing, or do we turn it down? Just the same, if lesser beings eventually collapse all human interaction, do we stay with lesser, or do we move toward necessary? Is this difficult?

A sustainable human future will watch as the human longings for collaboration meet in the middle with acquisitive self-interest. The two must and will bargain, having recognized the needs of the other. Models of collaboration roll efficiently along, while models of unchecked meritocracy are swept away every time. This negotiation between two extremes – acquisition and collaboration – will properly align what is now grossly misaligned. It's a matter of life and death for a lost, frightened, blameless, and paralyzed humankind.

EVOLUTIONARY DISCOMFORT/ORIGIN RESISTANCE

During *Homo sapiens'* 250,000-year migratory journey, they relied upon affiliative behavior. Our brain evolved in collaborative environments; we did not thrive in isolation. Modern brain research finally favors examination of the human brain and human behavior within the context of pack mentality. When we examine our ancient ancestors, we must commit to an immovable assumption — the notion that human beings evolved on Earth. That is a non-negotiable assumption. If evolutionary influence is not present at the heart of human sustainability efforts, no effort at all should be undertaken. Our beliefs will not save our species. A fusion of belief and knowledge is now possible.

The majority of humans are uncomfortable with the idea of evolution. We need not awaken Albert Schweitzer from his eternal slumber to determine why. This is my take. The human brain's first order of business is to interpret its environment, in the present moment. Pondering one's existence while operating heavy machinery is not a good idea — we can agree here. Pondering one's existence fell in for evolving humans long after the prime directive of immediacy was secured. We need straight answers in the present moment. If you disagree, close your eyes for ten minutes while actively moving around the room. You'll be in grave danger quickly, and things will get worse from there.

Let's inject the concept of evolution into this. Where are the straight answers we need? Natural selection is no less than an all-knowing mind. What's easier for you to fathom— God or the omniscient mind of natural selection?

Clearly, the spiritual default offers the most comfort, as quickly as possible. I don't recall ever knowing a person who routinely ruminates over the wonder of evolution. Where do we even start? Evolution is uncomfortable due to its complexity. Evolution has infinite implications. God is a handy concept when we're pressed for time, as we most certainly are, and increasingly so. Discomfort with evolution is not mysterious. We move on!

Specifically, research conducted across the world suggests that less than half of adults believe the world evolved. In 2015, ongoing surveys conducted by the Pew Research Center revealed a growing majority of people in the U.S. who believe in evolution. “Respondents were first asked whether ‘humans and other living things have evolved over time’ or ‘have existed in their present form since the beginning of time,’” the center notes in a report entitled *Americans, Politics and Science Issue.* “…In the most recent survey, 65% of adults say that humans and other living things have evolved, while 31% say humans and other living things have existed in their present form since the beginning of time.”

Origin discomfort is clearly an aggravating factor in problem-solving for human sustainability. If a sick patient rejects a medical diagnosis, it will often lead to a fatal outcome. By this analogy, so too can the rejection of human origin become fatal to human sustainability. Humanity must collectively own the reality of evolution or succumb to the delusions it chooses to explain its existence. Origin resistance robs an already put-upon humanity of self-diagnostic tools. It crowds out strategies for sustainability.

ESCAPIST BEHAVIOR

This can best be described as pleasure-seeking/pain-avoidance. The rapidly expanding gap between escapism and situational ownership is not manageable. Modern situational blindness is an event horizon that can never escape our gaze. Two things are equally true:

1. Situational ownership is a bummer.
2. Escapism is a killer.

NOËSIS

ESCAPISM

PLEASURE SEEKING/PAIN AVOIDANCE

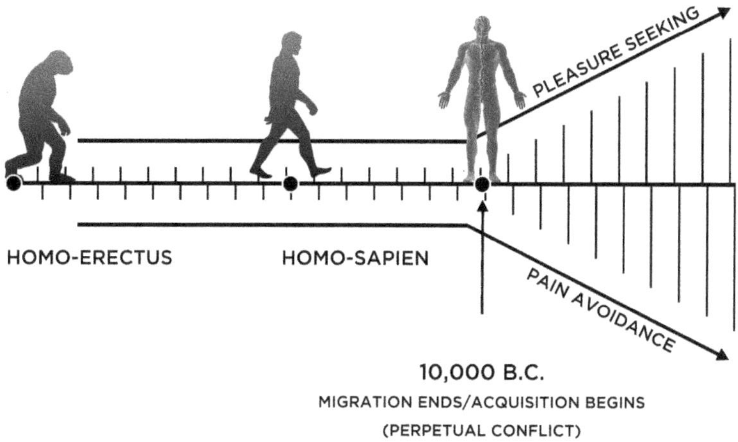

- HUMAN EXCESS IS GUARANTEED
- NO DNA-BASED BLUEPRINT
- NO ALTERING FORCE

INARGUABLE CONCLUSION:
COLLAPSE IS IMMINENT

SCAN THE QR CODE FOR A DETAILED EXPLANATION OF THIS ILLUSTRATION.

OUR CHOICE, the human choice, to survive or to perish, which will it be?

Statistics show that a missing child is increasingly endangered with every minute that passes. Again, the analogy is quite precise. A deluded humanity cannot be present. The ongoing normalization, the choice to simply look away, is suicidal. If we look away, we die! This inattentive blindness will summon commitment to extinction that Noēsis was built to describe and to actively avoid. Recall that once achieved, commitment is mathematically inescapable.

The point of commitment is knowable only in retrospect. Are we worried yet? False bravado among humans, the idea that we'll figure it out, combined with escapist behavior, invites the specter of catastrophic human failure. The now-swollen river of escapism will empty into an ocean of surrender. The children are watching. The children!

Where will surrender arrive? Where it always has! Societal Collapse.

What are we prepared TO DO?

Even though we stumbled into the evolutionary minefield of stationary and sedentary societies, there is a way out. If there were not, I wouldn't waste a second of your time.

To beat the mathematical odds and emerge as perhaps the single sentient species in the entire universe to survive technological infancy, we must slowly move away from failure.

To move away from failure, we must recognize failure. We stand together, all humans on Earth, on the metaphorical high dive. We'll need coaching to let go, to dive in, just as we did when we were children. Once we dove, hey, that wasn't so bad! Let's do it again.

As you exit, here are your takeaways. Blame is a threat to the human mind. We haven't figured that out. Blame and its forceful rejection are feeders of perpetual conflict. These are economic models and nothing more! They offer no solution, only certain extinction. Unnatural behavior pays, but we didn't know that it could kill. We need new clothes, ones that will allow all of us entry to a party called sustainability.

How can you help? Sit back and notice. Observe. See it differently. When you do, and if you stay with it, the anger will fall away like dead skin. A surviving humanity will look back and cringe, just like all of us today as individuals have cringeworthy memories. We didn't know that we missed a belt loop at the eighth-grade dance. We didn't mean to drop our tray in the lunchroom. And we humans are not to blame for even the slightest thing.

All together: It's not our fault! Let go of the self-loathing.

Blame is a liar. Blame creates and sustains perpetual conflict. Perpetual conflict is undefeated. We lacked the equipment, but no more. We see it. Remember?

It's the knowledge of our rarity that may provide some impetus to preserve ourselves. For a degrading humanity, any port in the storm. If these brutal descriptions look more like shock treatment than education, I remind the reader that shocking the heart saves lives every day. I make no apologies. I also remind anyone inconvenienced by these descriptions that inconvenience is not an argument.

Throughout this book, I've focused on the macro, the big picture. I've examined and I've documented the immensity of evolutionary weight on our present moments. I do not sensationalize extinction. What value does it bring? Instead, I warn of an existential sledgehammer called irrevocable commitment.

If you view me as the boy who cried wolf, you forgot to read to the end of that fable. The wolf arrives!

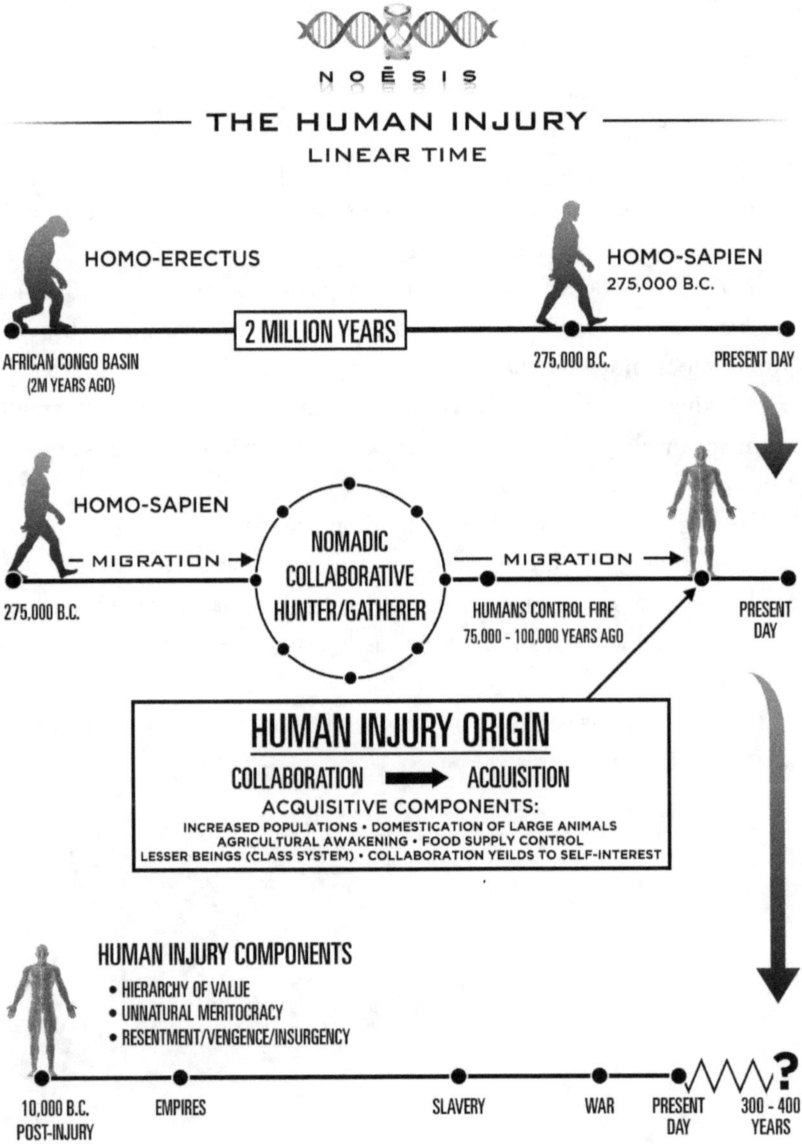
NOĒSIS
THE HUMAN INJURY
LINEAR TIME
HOMO-ERECTUS
HOMO-SAPIEN
275,000 B.C.
2 MILLION YEARS
AFRICAN CONGO BASIN
(2M YEARS AGO)
275,000 B.C.
PRESENT DAY
HOMO-SAPIEN
MIGRATION
NOMADIC
COLLABORATIVE
HUNTER/GATHERER
MIGRATION
275,000 B.C.
HUMANS CONTROL FIRE
75,000 - 100,000 YEARS AGO
PRESENT
DAY
HUMAN INJURY ORIGIN
COLLABORATION
ACQUISITION
ACQUISITIVE COMPONENTS:
INCREASED POPULATIONS • DOMESTICATION OF LARGE ANIMALS
AGRICULTURAL AWAKENING • FOOD SUPPLY CONTROL
LESSER BEINGS (CLASS SYSTEM) • COLLABORATION YEILDS TO SELF-INTEREST
HUMAN INJURY COMPONENTS
• HIERARCHY OF VALUE
• UNNATURAL MERITOCRACY
• RESENTMENT/VENGENCE/INSURGENCY
10,000 B.C.
POST-INJURY
EMPIRES
SLAVERY
WAR
PRESENT
DAY
300 - 400
YEARS
?

www.ingramcontent.com/pod-product-compliance
Lightning Source LLC
LaVergne TN
LVHW022041200926
848481LV00003B/729